AF322860

Arboviruses in Arthropod Cells In Vitro

Volume I

Editor

Conrad E. Yunker, Ph.D.

Cell Biologist
Hemoparasitic Diseases Research Unit
Agricultural Research Service
United States Department of Agriculture
and
Professor
Department of Veterinary Microbiology and Pathology
Washington State University
Pullman, Washington

CRC Press, Inc.
Boca Raton, Florida

Library of Congress Cataloging-in-Publication Data

Arboviruses in arthropod cells in vitro.

Includes bibliographies and index.
1. Arthropod-borne viruses. 2. Arthropod vectors.
3. Virus-vector relationships. 4. Cell culture.
I. Yunker, Conrad E. [DNLM: 1. Arboviruses--isolated &
purification. 2. Cells, Cultured. 3. Virus
Replication. QW 168.5.A7 A666]
QR201.A72A73   1986   616.9'2   86-4216
ISBN 0-8493-4504-9 (set)
ISBN 0-8493-4505-7 (vol. 1)
ISBN 0-8493-4506-5 (vol. 2)

© 1987 by CRC Press, Inc.

International Standard Book Number 0-8493-4504-9 (Set)
International Standard Book Number 0-8493-4505-7 (Volume I)
International Standard Book Number 0-8493-4506-5 (Volume II)

Library of Congress Card Number 86-4216
Printed in the United States

PREFACE

In the past decade, biology has exploded into new areas of technology with promises of benefits to immunology, genetics, medicine, and agriculture. Replicating in vitro systems of vector cells have been developed in the last 20 years that could aid substantially in the application of the new tools to solutions of problems in these disciplines. These systems of vector cells provide a useful means with which to study the biological interactions of arthropod-borne animal viruses (arboviruses) and cells of their intermediate hosts, and in addition offer a sensitive device for the isolation, identification, and characterization of arboviruses. These two volumes are intended to provide a comprehensive reference dealing with the status of arthropod cell cultures as systems for arbovirus growth. Beginning with an elaboration of the historical events that led to the development and virological application of arthropod tissue and cell cultures, chapters are presented that encompass methodology for cultivation of vector tissues and cells, characterization and identification of vector cell lines, and isolation, replication, and diagnosis of arboviruses in arthropod cell systems. Recent research findings concerning specific applications of vector cell systems to problems in infectious disease, both in the laboratory and in field situations (e.g., yellow fever and dengue viruses), are presented. Following this, the promise and limitations of arthropod cell systems as substrates for the production of diagnostic reagents and vaccines are discussed. Finally, the replicative dynamics of arboviruses in arthropod cell systems are reviewed. Included are discussions of persistence, fusion, host-range mutants, and the genetic influence of arthropod cells on virus growth.

The editor wishes to thank the numerous contributors to this volume, without whose cooperation the book would not have been possible. Special thanks are due to Charlotte A. Miller, whose careful attention to processing of the manuscripts is greatly appreciated and to Dr. Jane E. Homan, Dr. David Stiller, and Dr. Douglas M. Watts who kindly read parts of this book and offered their helpful comments.

Conrad E. Yunker

FOREWORD

Arthropod-borne viruses (arboviruses) are among the most versatile agents known. They replicate at ambient temperatures in their arthropod vectors and at much higher temperatures in their vertebrate hosts. They subsist meekly for long periods, sometimes years, in the invertebrate host. By contrast they may cause explosive disease in the vertebrate host, sometimes resulting in the death of the animal. The same effects hold in cell culture. The effect in arthropod cells is usually nonpathogenic, while that in vertebrate cells is often destructive. This moderation in the arthropod cell has long been a subject for intellectual discussion, but only recently a subject for serious scientific exploration resulting in surprisingly enlightening findings. These new advances are now encapsulated in this timely treatise.

There are well over 400 arboviruses. Among these are some of the earliest recognized viruses. Yellow fever in the family Flaviviridae was shown at the turn of the century by Walter Reed and his colleagues to be mosquito-borne. The early studies in *Aedes aegypti* established arthropods as major vectors of viruses. Also at the turn of the century scientists in South Africa described bluetongue of sheep and later African horsesickness caused by orbiviruses in the family Reoviridae. These arboviruses are major pathogens of domestic animals and are transmitted by *Culicoides* midges. In 1910, R. E. Montgomery isolated African swine fever virus, the only DNA-containing arbovirus. This tick-borne virus has spread from Africa to the Americas and now poses a serious economic threat to the swine industry. By 1925, the Indiana serotype of vesicular stomatitis virus of the family Rhabdoviridae was recognized as a disease agent of livestock. Still more mosquito-borne viruses were revealed in the 1930s: Rift Valley fever in the family Bunyaviridae, and Venezuelan, Eastern, and Western equine encephalomyelitis viruses of the family Togaviridae. By 1950, 35 arboviruses were known. These encompassed human and domestic animal pathogens of each of the major genera in the five largest families of arboviruses.

The rapid increase after 1950 in knowledge of arboviruses and their vertebrate hosts was accompanied by widespread experimentation in vertebrate cell cultures. The comparable studies with invertebrate cells were not to come for almost 2 more decades.

This lag is partly explained by the lack of easily propagated arthropod cells, but is also explained by the bandwagon effect in a major clan of virologists. These were molecular virologists making spectacular discoveries at a basic level using vertebrate cells with arbovirus models such as vesicular stomatitis, Semliki Forest, and Sindbis viruses. So much was happening so fast that few investigators paused to reflect that viruses such as vesicular stomatitis are basically arthropod viruses, not vertebrate animal viruses. An arbovirus infects its vertebrate host only briefly, just long enough to produce viremia so that the vector can be infected. The vector, in contrast, is infected for life, and indeed sometimes (as in the case of the sandfly with Indiana strain of vesicular stomatitis) has the potential to maintain the virus indefinitely by vertical transmission. It is ironic that much of the basic molecular virology knowledge was accumulated in the vertebrate cell which is only visited briefly in the life cycle of arboviruses in nature.

In the 1960s a handful of pioneers such as J. Peleg, T. D. C. Grace, J. Rehacek, K. R. P. Singh, K. Maramorosch, S. M. Buckley, M. G. R. Varma, M. Pudney, I. Schneider, and C. E. Yunker (with their collaborators) developed first mosquito, then tick cell lines, and just as importantly, they developed simplified media and propagation techniques. Later, A. Igarashi cloned the *Aedes albopictus* line of Singh, not only providing a tool with remarkably increased ability to produce progency virus, but also confirming the basic tenet that not all cells in a population are born equal. This single technology applied to dengue and other arboviruses has enabled the large-scale reliable replication so necessary for molecular studies.

This book tells for the first time in one volume, the complete story of both the theoretical and practical application of this recently developed methodology. We have answers (or at least the means to obtain the answers) for questions such as "How is an arbovirus able to replicate in cells so divergent as vertebrate and arthropod?", "What are the factors limiting the production of virus in arthropod cells?", and "What determines the specificity of a given arbovirus for a given arthropod cell?".

The arthropod cells have been a boon to those wishing to isolate arboviruses in nature. Since the cells are maintained at temperatures approaching ambient in tropical climates, the cells can be carried in the pocket, inoculated in the field, and sent to the laboratory without special handling. It is no longer necessary to maintain a cold-chain for shipping field specimens suspected of containing an arbovirus. In most cases the arthropod cell, once infected, remains persistently infected, maintaining the culture of virus for weeks if necessary.

An unexpected dividend is now resulting from the use of mosquito cells for primary isolation of viruses from mosquitoes. A large number of viruses has recently been isolated which previously went unrecognized on inoculation of vertebrate cell cultures or animals. Most of these agents are not yet identified. They may well be viruses indigenous to mosquitoes and of no public health consequence, but the parallel is drawn to the isolation in the 1950s and 1960s of large numbers of new viruses from arthropods in baby mice. Some of these such as Crimean-Congo hemorrhagic fever and Oropouche viruses were initially thought to be of minor public health importance, but were subsequently shown to be major human pathogens. The human exposure to arthropods and to arthropod cells (some presumably infected with these viruses) can be a daily event, especially in the tropics, and the potential of the new mosquito cell isolates should not be disregarded.

The development and use of tick cells in culture have not come as easily or as quickly as the use of mosquito cells. As the technology progresses, however, this field will inevitably blossom and offer new vistas for study. The young investigator should take special note of tick cell culture as an area to be exploited.

Dr. Yunker is to be congratulated on bringing together the information on arthropod cells and arboviruses. There is yet much to be done and learned in this fascinating system. These two volumes will serve as a catalyst at this time of rapidly advancing technology.

Robert E. Shope, M. D.
Director
Yale Arbovirus Research Unit
Department of Epidemiology and
 Public Health
School of Medicine
Yale University
New Haven, Connecticut

CONTRIBUTORS

Danielle Blondel
Scientific Researcher
Laboratoire de Génétique des Virus
Centre National de la Recherche
 Scientifique
Gif-Sur-Yvette, France

Francoise Bras, D.Sc.
Assistant Chief
Laboratoire de Génétique des Virus
Centre National de la Recherche
 Scientifique
Gif-Sur-Yvette, France

Susan E. Brown, Ph.D.
Associate Research Scientist
Department of Epidemiology and
 Public Health
School of Medicine
Yale University
New Haven, Connecticut

D. Contamine, Dr.Sc.
Scientific Researcher
Laboratoire de Génétique des Virus
Centre National de la Recherche
 Scientifique
Gif-Sur-Yvette, France

Sybille Dezélée, D.Sc.
Scientific Researcher
Laboratoire de Génétique des Virus
Centre National de la Recherche
 Scientifique
Gif-Sur-Yvette, France

Kenneth H. Eckels, Ph.D.
Assistant Chief
Department of Biologics Research
Walter Reed Army Institute of
 Research
Washington, DC

Peter-Joachim Enzmann, Dr.rer.nat.
Wissenschaftlicher Oberrat
Federal Research Centre for Virus
 Diseases of Animals
Tübingen, West Germany

Duane J. Gubler, Sc.D.
Chief
Dengue Branch and
Director
San Juan Laboratories
Center for Infectious Diseases
Division of Vector-Borne Viral Diseases
Centers for Disease Control
San Juan, Puerto Rico

Akira Igarashi, M.D.
Professor and Chairman
Department of Virology
Institute for Tropical Medicine
Nagasaki University
Nagasaki, Japan

Martine Jozan, M.D., D.R.P.H.
Research Virologist
Adjunct Lecturer
School of Public Health
University of California at Los Angeles
Los Angeles, California

Christoph Kempf, Ph.D.
Institute of Hygiene and Microbiology
University of Bern
Bern, Switzerland

Dennis L. Knudson, D. Phil.
Associate Professor of Epidemiology
Department of Epidemiology and
 Public Health
Yale University
New Haven, Connecticut

Hans Koblet, M.D.
Professor
Institute of Hygiene and Microbiology
University of Bern
Bern, Switzerland

Goro Kuno, Ph.D.
Research Microbiologist
San Juan Laboratories
Division of Vector-Borne Viral Diseases
Centers for Disease Control
San Juan, Puerto Rico

Christine Kurstak, M.D., Ph.D.
Head
Laboratory of Virology
Hospital Hôtel-Dieu de Montréal
University of Montreal
Montreal, Québec, Canada

Edouard Kurstak, Dr. d'Etat, Ph.D.
Professor and Director
Department of Microbiology and
  Immunology
Faculty of Medicine
University of Montreal
Montreal, Québec, Canada

Colin J. Leake, Ph.D.
Lecturer
Department of Entomology
London School of Hygiene and
  Tropical Medicine
London, England

Helmut Mahnel, Dr. med. vet.
Professor
Faculty of Veterinary Medicine
University of Munich
Munich, West Germany

Eberhard Munz, Dr. med. vet.
Professor
Faculty of Veterinary Medicine
University of Munich
Munich, West Germany

Adames Omar, Ph.D.
Institute of Hygiene and Microbiology
University of Bern
Bern, Switzerland

Anne-Marie Petitjean
Ingeneer
Laboratoire de Génétique des Virus
Centre National de la Recherche
  Scientifique
Gif-Sur-Yvette, France

Mary Pudney, Ph.D.
Research Scientist
Department of Biochemical
  Microbiology
Wellcome Research Laboratories
Beckenham, Kent, England

Josef Reháček, Ph.D., D.Sc.
Research Scientist
Department of Rickettsiae
Institute of Virology
Slovak Academy of Sciences
Bratislava, Czechoslovakia

Max Reimann, Techn. Assist.
Faculty of Veterinary Medicine
University of Munich
Munich, West Germany

Imogene Schneider, Ph.D.
Department of Entomology
Walter Reed Army Institute of
  Research
Washington, DC

Robert McNair Scott, M.D., F.A.A.P.*
Head
Department of Virology
NAMRU-3
FPO New York, New York

Victor Stollar, M.D.
Professor
Department of Molecular Genetics and
  Microbiology
Robert Wood Johnson Medical School
University of Medicine & Dentistry of
  New Jersey
Piscataway, New Jersey

Danielle Teninges, D.Sc.
Scientific Researcher
Laboratoire de Génétique des Virus
Centre National de la Recherche
  Scientifique
Gif-Sur-Yvette, France

* Present affiliation: Assistant to Director, Division of Communicable Disease and Immunology, Walter
  Reed Army Institute of Research, Washington, D.C.

Peter Tijssen, Ph.D.
Research Associate
Institut Armand-Frappier
Laval des Rapides,
Québec, Canada

M. G. R. Varma, Ph.D., D.Sc.
Professor of Medical Entomology
Department of Entomology
London School of Hygiene and
 Tropical Medicine
London, England

Francoise Wyers, D.Sc.
Scientific Researcher
Laboratoire de Génétique des Virus
Centre National de la Recherche
 Scientifique
Gif-Sur-Yvette, France

To Samira

TABLE OF CONTENTS

Volume I

Volume II

APPLICATIONS OF ARTHROPOD CELL CULTURES TO DENGUE
VIRUS IDENTIFICATION AND VACCINE PRODUCTION

REPLICATIVE DYNAMICS OF ARBOVIRUSES IN ARTHROPOD
CELL SYSTEMS

Introduction 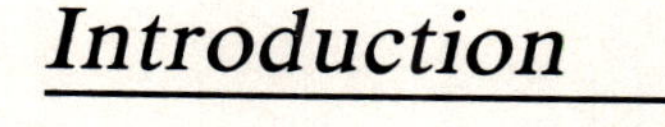

Chapter 1

OF ARBOVIRUSES, ARTHROPODS, AND ARTHROPOD CELL CULTURES: HISTORY AND EXPECTATIONS

M. Jozan

TABLE OF CONTENTS

"There are two stumbling blocks in the study of natural history, the first is not to have a method, and the second to relate everything to a particular system; one can use an already established method as a commodity for studying; it should be viewed as a facility for communicating . . . ."

Buffon

I. INTRODUCTION

The Yellow Fever Saga which occurred in 1901 introduced a new class of viruses and initiated novel concepts regarding transmission mechanisms of viral agents. The so-called Arthropod-borne viruses (arboviruses) replicate within a specific arthropod vector which remains infected for its lifetime without any apparent deleterious effects. Following an extrinsic incubation period of assembly and replication in its vector, the virus is transmitted by bite to a susceptible animal host which can become a long-term reservoir. The primary maintenance cycle of arboviruses is zoonotic and man is an incidental intruder and temporary host who may develop a variety of mild to severe clinical infections.

Yellow-fever studies also provided modern arbovirologists with one of the most useful tools for isolation and epidemiologic investigations, namely white mice which, in 1930, replaced the rhesus monkey as an experimental animal. Suckling mice are susceptible to most arboviruses, but this in vivo system is not always available and/or affordable, hence the need for developing the less cumbersome and space-dependent tissue culture system was justified.

II. DEVELOPMENT OF TISSUE CULTURE TECHNIQUES FOR THE STUDY OF ARBOVIRUSES

The Maitland type of tissue culture had already been used in 1936 by Lloyd, Theiler, and Ricci who established the wild strain of yellow fever (YF) in a culture medium containing minced mouse embryo tissues.[1] Survival and serial passage of the virus was possible but there were no means of judging the quality of cell-virus interaction, and thus controlling the multiplicity of infection. This became possible with the advent of methods whereby cells were freed from tissues by enzymatic digestion and allowed to grow as a single monolayer on a glass substrate, later to be replaced by plastic. The technique was further refined by Dulbecco who demonstrated the formation of plaques with Western equine encephalitis virus (WEE) in chick embryo cell monolayers overlaid with a semisolid nutritive medium.[2] The plaques or foci of local destruction were considered to be produced by progeny of a single virus particle. By 1956 the plaque technique was successfully used by Dulbecco and his colleagues[3] to demonstrate the kinetics of WEE virus replication and its neutralization by hyperimmune horse and rabbit antisera.

Baseline studies of arbovirus growth in a variety of cell-culture systems were undertaken at about the same time in four different locations. The Rockefeller Foundation Laboratories in New York and abroad had shifted their interest and support from the study solely of YF to that of all known arboviruses. Thus in Cairo, Egypt, as a consultant to NAMRU-3, Frothingham demonstrated the selective growth of Sindbis virus (SIN) in cell cultures of chick embryos, human uterine tissues, monkey testicles, and even the continuous HELA cell line.[4] It is noteworthy that at the time no cytopathic effect (CPE) was obtained in HELA cells. In 1956 at the Virus Research Center in Poona, India, a successful plaque assay in primary chick embryo and rhesus monkey cells was developed by Bhatt and Work for the study of West Nile (WN) and Japanese

B encephalitis (JBE) viruses.[5] This was the prelude to a long tradition of tissue culture research which eventually culminated with the establishment by Singh of invaluable invertebrate cell lines.

Meanwhile, Buckley had been recruited in 1956 by the New York laboratory with the task of introducing tissue culture techniques applicable to the isolation and identification of arboviruses. Braving her isolation and the suspicion of some toward the usefulness of novel methods, she undertook systematic studies of classified arboviruses in primary chick embryo cells and in the HELA cell line established by Gey and co-workers in 1952. The scope of cell sensitivity was assessed and for each virus the range of CPE and growth curves were described. Progressively such identification techniques as sensitivity to solvents and neutralization tests routinely performed in mice were adapted to tissue culture and the foundations were laid for all subsequent work in this system.[6]

At the same time the Communicable Disease Center of the U.S. Public Health Service in Atlanta, Ga. was not idle. Comparative studies of arbovirus infections of mice and tissue culture showed that primary cultures of hamster kidney cells exhibited a virus sensitivity similar to that of suckling mice with, in fact, the occurrence of CPE preceeding that of death in mice.[7] This method was to be used extensively by Reeves and colleagues for virus isolations during the following epidemics of Western equine encephalitis (WEE) in California.

III. EARLY STUDIES OF ARBOVIRUS REPLICATION IN THEIR VECTORS

The refinement of primary cell cultures and the proliferation of mammalian cell lines nurtured multiple studies of arbovirus replication for which by far, Sindbis (SIN) became the choice reagent and a standard model for molecular virologists. Despite its usefulness and practicality, vertebrate cell culture was not the ultimate answer and furthermore, because of the nature of arboviruses themselves, it was logical to believe that broader insight into their replication would be provided by laboratory studies of the vector or cells derived from their tissues.

Indeed as early as 1932, Davies et al. showed that transmission of the Asibi strain of YF by *Aedes aegypti* was dependent upon the temperature at which the mosquitoes were reared.[8] Transmission did not occur at 17 to 19°C, although the virus was present in mosquito tissues; moreover, transmission could be restored when infected mosquitoes were transferred for 6 days at room temperature. Similar results were obtained in laboratory studies of the *Saimiri*-monkey-*Haemagogous* mosquito cycle of jungle YF.[9] Comparison of YF replication in *Culex thalassius* and *Aedes aegypti* introduced the notion that the quality of virus transmission was dependent upon the vector and virus strain as well.[10] Davies and Hurlbutt, independently studying the competence of various mosquito vectors for West Nile infection, were the first to point out that a certain multiplicity of infection is necessary to achieve successful transmission.[11-13] Infection of *Culex pipiens* with JBE by Lamotte[14] demonstrated that within 10 min following a blood meal, virus could be found in hemolymph; migration to the midgut preceded invasion of the ovaries and all other tissues by the fourth week. This rapid digestion of an infected meal by mosquitoes contrasted with the slow adsorption of tick-borne encephalitis virus in ticks described by Pavlovsky in 1945.[13,14]

Experimental infection of mosquitoes or ticks with arboviruses thus provides a means of studying the kinetics of their replication and establishing the basic concepts of transmission mechanisms, yet the techniques require the costly and time consuming maintenance of an insect colony and, besides, success in rearing some species is not

Table 1

BASIC DIFFERENCES BETWEEN INSECT HEMOLYMPH AND
VERTEBRATE BLOOD AS DETERMINANTS FOR DESIGN OF
INVERTEBRATE TISSUE CULTURE MEDIA[a]

Substance	Insect hemolymph (mg/100ml)	Vertebrate blood (mg/100ml)
Inorganic		
Mole ratio Na + K/Ca + Mg	5—10	10—25
	Mole ratio Na/K: 0.1—25	
	<1 in phytophagous insects	
	>1 in zoophagous insects	
Organic		
Proteins	100—5,000	20,000
	Many with enzymatic activity	
Carbohydrates	2,000	90
	Trehalose specific to insects yet of no essential benefit in cell culture	
Amino acids	200—2,000	60
	Mostly of L configuration	Mostly of D configuration
	Account of 40% of osmotic pressure	Account for 1% of osmotic pressure
Vitamins	B Group essential may be provided by yeasteolate and lactal bumin	High content and very diverse
Fatty acids — Sterols	Incorporated into glycerides, phospholipids, and Sterolesters	Importance of linoleic and arachinodic acid for maintenance of cell membrane
	No Choline synthesis	
Sugar phosphates	50	30
Nitrogenous waste	10	30

[a] Data compiled from Bursell, E., *An Introduction to Insect Physiology,* Academic Press, 1970, and *The Growth and Requirements of Vertebrate Cells In Vitro,* Waymouth, Ch., Ed., Cambridge University Press, 1981.

assured. The switch from live mosquitoes to the culture of their tissues was the next logical step.

IV. EMERGENCE OF INVERTEBRATE CELL CULTURE METHODS

The trials and tribulations of invertebrate cell culture methodology have been reviewed in detail.[15-18] Briefly, essential differences between vertebrate blood and insect hemolymph (Table 1) prompted the belief that the latter had to be the vital component of any nutritive medium in order to achieve successful growth of insect cells. Heat-treated hemolymph, in fact, provided a large molecule fraction which did facilitate migration, but had little direct effect on growth promotion.[19] Other substitutes, such as chick embryo extracts, were proposed, but it was the ultimate use of fetal calf serum and adaptation to such media as Eagle's or Leibovitz's L-15 which gave a final impulse to the development of mosquito cell lines in particular. On the other hand, primary tick tissue cultures and later on continuous cell lines were grown mostly in media designed for vertebrate cell cultures. A variety of tissues may be grown from arthropods, but embryonic, larval, and developing adult tissues are preferable. Primary tissue cultures require a large number of arthropods of which some are only seasonally available and are often doomed by contamination and/or limited growth which renders their use quite aleatory for long-term studies of virus replication. The near simultaneous establishment of mosquito cell lines by Grace[20] and Singh[21] set the pace for basic virus research and comparative studies.

V. GENERAL CHARACTERISTICS OF ARTHROPOD CELL CULTURES

Mosquito cell cultures and to a certain extent tick cell cultures are remarkable for their heterogeneity; pseudoepitheliocytes, fibroblasts, and amoebocytes are found and the morphology of these undifferentiated cells is similar to that of vertebrate cultured cells. Yet some traits are particular to invertebrate cultures, namely: (1) a natural tendency to display multinucleate cells, (2) a marked ability for heterotypic exclusion, (3) a propensity to develop into diploid/haploid lines, and (4) optimum growth at temperatures below 30°C and at acid pH.

VI. ARTHROPOD CELL CULTURE AND SUSCEPTIBILITY TO ARBOVIRUS INFECTION

There are at this point in time more than 170 continuous cell lines from arthropods representing 99 species of Diptera, 7 species of Acari (Ixodidae), and 36 species of Lepidoptera;[22-24] some of these used in arbovirus studies are listed in Table 2. Many sublines or clones are omitted since their importance and advantages over the original line remain to be assesed.

The susceptibility of one or more mosquito cell lines to infection with arboviruses has already been well described[46-48] and may warrant only the following comments: The final impetus to the exploitation of arthropod cell cultures was given by the studies of JBE and other viruses in the *Antheraeae eucalipti* and *Aedes aegypti* cell lines of Grace.[49,50] These studies followed early recognition[51] that WEE could be successfully grown in *Aedes aegypti* larvae cells as well as in other tissues from this mosquito in which Eastern equine encephalitis (EEE) and WN could multiply.[52,53] Visualization by immunofluorescence of tick-borne encephalitis (TBE) viral antigen in tick hemocytes was at the same time demonstrated by Rehacek and Maerova.[54]

The general characteristics of invertebrate cell culture for arbovirus studies can be summarized as follows:

1. The successful growth of arboviruses in dipteran cell lines is seldom accompanied by CPE and, if present, CPE may range from limited structural changes to formation of syncytia or total cell lysis. No CPE has been described so far in tick cell lines that support the multiplication of viruses.
2. In the absence of CPE, detection of virus multiplication must rely upon the inoculation of infected cells and/or medium into mammalian cell lines or into suckling mice in which either CPE or death will be observed; or the viral antigen can be visualized in infected cells by immunofluorescent or immunoenzymatic methods; in some cases a hemagglutinating and/or complement-fixing antigen is produced.
3. Plaques can be obtained in some mosquito cell lines whether or not arboviruses do produce a CPE.[55,56]
4. The growth of arboviruses within the 5 taxons of their classification will differ according to the species of cell line and for a given cell line, with each virus type; the epidemiologic implications of this characteristic is discussed below.
5. Induction of virus persistence in mosquito cell lines is possible with most arboviruses capable of replicating in this system.

Table 2
CONTINUOUS
ARTHROPOD CELL LINES
SUSCEPTIBLE TO ONE OR
MORE ARBOVIRUSES

Cell line	Ref.
Lepidoptera	
Antheraea eucalypta	25
Diptera	
Aedes	
A. aegypti	20
A. albopictus	21
A. dorsalis	26
A. malayensis	27
A. novalbopictus	28
A. pseudoscutellaris	27
A. taeniorhyncus	109
A. vexans	29
A. w-albus	30
A. vitattus	31
Anopheles	
A. gambiae	32
A. stephensi	33
Culex	
C. molestus	34
C. quinquefasciatus	35
C. salinarius	36
C. tarsalis	37
C. tritaeniorhyncus	38
Toxorhynchites amboinensis	40
Acari	
Ixodidae	
Boophilus microplus	41
Dermacentor parumapertus	42
D. variabilis	43
Haemaphysalis obesa	44
H. spinigera	44
Rhipicephalus appendiculatus	45
R. sanguineus	44

It is worth mentioning that comparison of different cell lines for their susceptibility to arbovirus infection is difficult indeed since there is no standard protocol for the choice of line, circumstances of cultivation, and selection of virus strain. Therefore for some viruses as well as for some cell lines, information is too sparse to be of any significant use.

As with mammalian tissue cultures, any justification for the use of arthropod cell lines depends upon answers to the following questions:

1.   Are arthropod cell cultures superior to their vertebrate counterpart for virus isolation and characterization?

2.   What can be gained from the study of arbovirus replication in arthropod cell cultures which could be obtained by other methods?

3.   What is the significance of virus persistence and what bearing does it have on our understanding of transmission mechanisms?

4. How applicable to an understanding of infection in nature is the study of virus infection of arthropod cells?

VII. USE OF ARTHROPOD CELL CULTURES FOR PRIMARY ISOLATION OF ARBOVIRUSES FROM FIELD-COLLECTED SPECIMENS

Our understanding of transmission and causation of arboviral disease relies primarily upon proper isolation and identification of a virus from arthropod vectors, reservoirs, and patients. Members of the genus of *Alphavirus* can be recovered equally well in mice and vertebrate or invertebrate cell systems, but such is not the case for some flaviviruses, particularly the mosquito-borne dengue (DEN), St. Louis encephalitis (SLE), and Murray Valley encephalitis (MVE) viruses. The recovery of a virus in any system does not automatically warrent its sensitivity as demonstrated by Work and Jozan.[108] The comparison of virus recovery from BHK/PS cells and suckling mice, respectively inoculated with 836 and 625 pools of mosquitoes collected on the same field site and at the same time, showed that (1) the isolation rate for WEE was comparable in both systems, whereas recovery of SLE was 4 to 6 times lower in tissue culture and (2) this was confirmed when 65 known positive pools were reinoculated into both systems with a recovery rate of 83% for WEE, 47% for SLE, and nil for California encephalitis virus (CE).

The difficulty of recovering dengue viruses from human sera and other field specimens in vertebrate cell cultures as well as in suckling mice is well documented and was instrumental in the development by Rosen of the mosquito intrathoracic inoculation technique. The latter provides a mean of virus amplification in arthropods following a specific period of incubation. Direct characterization of infected cells can be achieved by immunofluorescent or immunoenzymatic staining of head squashes.

A. Results of Virus Isolation Attempts in Arthropod Cell Culture

Primary explants from mosquito larvae have been used for the isolation of arboviruses.[51,52] These small scale experiments were quickly displaced in favor of the more convenient and reliable use of continuous cell lines. Table 3 summarizes experimental conditions and results from ten different isolation studies and shows that:

1. It is possible to obtain CPE in Singh's *Aedes albopictus* cells with the four types of DEN virus.
2. Yellow fever virus may be isolated in the *Aedes pseudoscutellaris* line where it produces very early CPE with total lysis.
3. *Aedes albopictus* cells are more sensitive than either mice, Vero cells, or LLCMK2 cells for the isolation of DEN, but are slightly less sensitive than the intrathoracic inoculation technique.
4. Infected mosquito cells can produce a complement-fixing (CF) antigen which may be detected as early as 7 days for DEN-1 and 2, but in the case of closely related strains, the CF antigens are cross-reactive; production of a CF antigen may also be inconsistent.
5. Infected mosquito cells can produce a hemagglutinin,[64] but the titer is generally low and sometimes natural hemagglutinins can be found in control cells.
6. When there is no visible CPE and little or no detectable CF antigen, characterization of the isolate must rest upon innoculation into a susceptible mammalian cell line such as Vero, LLCMK2, or PS, or upon direct staining of infected cells by immunofluorescence using conventional hyperimmune sera or, at best, monoclonal antibodies.[66,67]

Table 3
COMPARATIVE RESULTS OF VIRUS ISOLATION ATTEMPTS FROM FIELD SPECIMENS IN DIFFERENT IN VIVO AND IN VITRO SYSTEMS

Virus study	Material	Isolation systems and results		Main observations	Ref.
Dengue	Human sera (HS)	*Aedes albopictus* Vero	CF antigen with tissue culture fluid	CF appears earlier in mosquito cells CF appearance different for each dengue type Cross reactivity of CF at day 9	61
Dengue	25 Human sera (HS)	*A. albopictus* Vero Mice	=22 Positive =18 Positive =Negative but resist challenge	Cytopathic effect with complete lysis (Dengue 1 and 2) Progressive syncitial formation (Dengue 3 and 4)	58
Dengue 2	116 Human sera (HS) 22 Mosquito pools (MP)	*A. albopictus* LLCMK2 Mice	=29 HS, 5 MP positive =20 HS, 3 MP positive =25 HS, 4 MP positive	Poor CPE in LLCMK2 Syncitia in *A. albopictus* Identification by plaque neutralization in LLCMK2	59
Dengue	664 Human sera (HS)	*A. pseudoscutellaris*	=238 Positive =218 at first pasage	Identification by CF CF cross reactive not always improved with Freon	62
	25 Known isolates (HS)	*A. pseudoscutellaris* Intrathoracid mosquito inoculation LLCMK2 Mice	=Positive =Positive =Negative =Negative	Percent of isolations decreases as HI titer increases Intrathoracic inoculation of mosquitoes more sensitive than mosquito cells	
Dengue	16 Dengue isolates serially passed in either *Ae. albo.* or *T. amboinensis*	*A. albopictus* C6/36 *A. pseudoscutellaris*		No CPE in either mosquito cells Indentification by CF or FA Higher percent of infected cells; shorter incubation period at 32°C	66
Dengue	84 Mosquito isolates	*A. albopictus* C6/36 LLCMK2	CPE CPE and plaques	No CPE with Dengue 4 In both cells CPE at 36 hr with	64

					Ref.
	3 Human sera (HS)			Dengue 2 Virus titer higher in C6/36 than in LLCMK2 In C6/36 simultaneous appearance of hemagglutinin and virus Ability of monoclonal FA antibody to distinguish between strains	
Yellow fever	5 Mosquito pools 2 Human livers 1 Sentinel rhesus	*A. pseudoscutellaris*	CPE	CPE as soon as day 4 with complete lysis Very low titer with liver isolates, high with rhesus	60
Yellow fever	5 Mosquito pools 2 Human livers 1 Sentinel rhesus	*A. albopictus*	CF with tissue fluid	Inconsistent CF antigen Identification in Vero Vero more sensitive than *A. albopictus* cells	63
St. Louis encephalitis	1 Mosquito pool 1 Fox isolate	*A. albopictus* (Singh) *A. aegypti* (Peleg)	CPE No CPE	Multiplication in *A. aegypti* without CPE Identification in PS cells	65
St. Louis encephalitis	25 Mosquito pools Isolates	*A. albopictus*	22 Recovered between days 6 to 8	Very rapid identification by IFFA (72—96 hr) Plaque neutralization in Vero (2 to 3 weeks)	67

Note: CF is complement fixation, CPE is cytopathic effect, FA is fluorescent antibody, IFFA is immunofluorescent focus assay, and HI is hemagglutination inhibition.

B. Factors Influencing the Growth of Arboviruses in Arthropod Cell Cultures

The failure to promote virus growth in cell cultures may be attributed in general to such factors as temperature, pH, multiplicity of infection, virus strain, cell type, and composition of medium.

1. Influence of Temperature

The role of temperature as a critical event in the development of viral functions was described for poliovirus in the 1960s. For each virus there is an optimum temperature of successful replication above or below which multiplication will be reduced or inhibited.[68] Thermoresistant mutants or variants may emerge at temperatures above optimal and will sometimes show a greater virulence; at temperatures below optimal, thermosensitive variants/mutants are often characteristically attenuated.

The very fact that the preferential growth of arthropod cells is done at 28 and 30°C may hamper successful recovery of viruses. The influence of temperature upon the production and quality of plaques has been demonstrated for alphaviruses in vertebrate cell cultures.[69] The growth of Venezuelan equine encephalitis virus (VEE) in minced *Aedes aegypti* larvae was similarly affected by temperature, since optimum virus adsorption would occur between 4 to 30°C, whereas at 37°C a decrease in adsorption rate and little subsequent virus replication were observed.[70] Comparative studies of the EEE virus in similar systems did not reveal a temperature dependency. When virus growth in both vertebrate and invertebrate cells is compared, the peak titer is seen to occur at a later time in the latter, probably due to an increased latent period determined by the temperature of incubation. As shown for Semliki Forest virus (SFV), the shortest latent period is directly related to the optimum temperature of cell growth.[71] growth.[71]

2. Multiplicity of Infection

Inoculation of a given amount of virus onto a cell monolayer does not insure its total adsorption. Between inoculation and actual entry into the cells there is a loss of virus due to its inability to contact the cells, adsorption to cellular debris, inactivation, or loss by aggregation. As noted by Johnson,[70] below a certain multiplicity of infection, little or no virus is produced in the culture. Similar observations with different alpha- and flaviviruses were made in mosquito cells,[13,14] but differences among virus strains as well as cell lines used preclude comparison.

3. Cell Type and Circumstances of Culture

The concept that each cell line has a defined susceptibility for each virus or group of viruses is not new, but appears to be more critical in arthropod cells. For instance the length of latent period for CE virus was different in *Aedes aegypti* and *A. albopictus* cell lines;[72] similar results were reported for growth of Kunjin (KUN) virus in *A. albopictus,* Vero, and PK cell lines.

The susceptibility to infection with some arboviruses will also vary with the transfer level of the cells or with the clone. Such is the case with cloned variants of Singh's *A. albopictus* line (clone C6/36 and others) which may accomodate the growth of SIN virus with production of CPE[73] or of Chikungunya (CHIK) as well as JBE[74] and the four types of DEN virus.[75] Such differences may relate to the disappearance of most cell proteins at the time of maximum virus replication and release.[76,77] Similarly, in *A. albopictus* cells and clone C6/36 infected with KUN virus, the rate of protein synthesis was shown to be slower in the former line, and no hemagglutinin was detected as compared to the C6/36 clone.[80] Furthermore, when 0.1% bovalbumin was substituted for fetal calf serum, the virus yield decreased, which points to the importance of medium composition.

4. Contamination

Whereas bacterial and mycoplasma contamination may rapidly erase any hope of a decent and workable tissue culture system, latent infections of invertebrate cells with either an arbovirus or an insect virus are far worse.[79-80] Such contamination, if undetected, may interfere with the otherwise successful growth of arbovirus isolates.

To conclude, dipteran cell lines can usefully replace suckling mice or mosquito inoculations when such systems are not available provided that conditions of culture and the kind of cell are carefully chosen for optimum virus growth. A definite advantage of arthropod cell cultures over their vertebrate counterparts is their ability to grow at low temperatures and be maintained for long periods of time with little or no changes of medium. This makes the transport, storage, and inoculation of cultures possible under field conditions, thus providing a better chance of virus recovery.

VIII. ARTHROPOD CELL CULTURES FOR IDENTIFICATION OF ARBOVIRUSES

An ultimate goal of virus characterization is to define an array of traits which can be used to determine their phylogenic appurtenances and so perceive their origin and evolution. Information on virus replication drawn from in vitro studies does not necessarily reflect the situation in nature because such studies are the result of experimental manipulation performed in a carefully controlled environment. In the case of arboviruses, arthropod cell cultures (more than any other system) may in fact provide information as to some mechanisms of transmission and maintenance. Table 4 compares results obtained by different investigators with one or more arthropod cell line; the few viruses presented here were the most commonly studied and belong to four of the five recognized arboviral taxons. As shown, the growth of some viruses may be restricted to a given species or cell type. When the susceptibility of arthropod cells is considered for each virus in relation to its primary or probable vector species in vivo, certain facts emerge:

1. Mosquito-borne arboviruses will consistently grow in dipteran cell lines[81-83] and some grow in primary or continuous tick-tissue culture
2. As shown for O'Nyong Nyong (ONN) and CHIK viruses, optimum growth is better achieved in cell cultures corresponding to the primary in vivo vector whereas poor or no growth occurs in cells from other species[84,85]
3. Mosquito-borne arboviruses which incidentally have also been isolated from ticks, will replicate in tick cells (WN, SFV)
4. Tick-borne viruses belonging to all taxons will not constantly replicate in mosquito cells[86,87] but will successfully grow in ixodid tick cells without production of CPE. Of particular interest is the observation that Punta Salinas, a putative member of the genus *Nairovirus* isolated from argasid ticks, will replicate only poorly in ixodid cell lines.[88]

The differential growth patterns of a virus isolate in a battery of selected arthropod cells from different species may thus provide some clues of epidemiologic importance.

IX. PERSISTENCE

Whereas persistent infection of vertebrate cell lines has been laboriously achieved with certain arboviruses, it is a prominent feature of mosquito cell lines (and probably those of other arthropods) to favor the establishment and maintenance of such infections possibly with any arbovirus. Because of their morphologic heterogeneity, arthro-

Table 4
GROWTH OF SELECTED ARBOVIRUSES IN CONTINUOUS MOSQUITO AND TICK CELL LINES

		Dipteran cell lines					Ixodid cell lines		
Taxon/Genus/Virus[a]	Known vectors	*Aedes albopictus*	*A. aegypti*	*A. pseudo-scutellaris*	*Anopheles stephensi*	*Culex quinque-fasciatus*	*Rhipicephalus appendiculatus*	*Boophilus microplus*	*Dermacentor variabilis*
Togaviridae									
Alphavirus									
Chikungunya	Cu/An	+/CPE[b]	+	+	Slow	NG	ND	ND	+
Semliki Forest	Cu/An/T	+/CPE[b]	+	+	NG	ND	+	ND	ND
Sindbis	Cu/An	+/CPE[b]	+	+	++	ND	INC	ND	ND
O Nyong Nyong	An	+	+	+	++	ND	Poor	ND	+
Flavivirus									
Japanese B encephalitis	Cu/An	+/CPE	+	+	+	+	NG	ND	ND
Dengue	Cu	+/CPE	+	+	+	+D2	ND	ND	ND
West Nile	Cu/An/T	+/CPE	+	+	ND	++	+	ND	ND
Louping I11	Ix	NG	NG	NG	ND	ND	ND	+	ND
Langat	Ix	NG	NG	NG	ND	ND	+	+	+
Bunyaviridae									
Bunyavirus									
California encephalitis	Cu	+	+	+	ND	+	NG	NG	ND
Germiston	Cu	+	+	+	ND	+	NG	NG	ND
Tahyna	Cu/An	+	+	+	+	ND	NG	NG	ND
Nairovirus									
Zirga	Ix	NG	NG	NG	ND	ND	+	ND	ND
Quaranfil	Ix	NG	NG	NG	ND	ND	+	ND	ND
Lanjan	Ix	NG	NG	NG	ND	ND	+	ND	ND
Punta Salinas	Ar	ND	ND	ND	ND	ND	Slow	ND	ND
Rhabdoviridae									
Vesiculovirus									
Chandipura	Ph	+	+	NG	+	ND	NG	NG	NG
Lyssavirus									
Otonkan	Unk	+	NG	ND	ND	ND	ND	ND	ND

Kotonkan	Unk	+	NG	NG	ND	ND	ND	ND	ND
Mokola	Unk	+	NG	ND	ND	ND	ND	ND	ND
Reoviridae									
Orbivirus									
Colorado Tick fever	Ar/Ix	+	NG	NG	ND	ND	ND	ND	++
Kemerovo	Ar/Ix	Poor	+	NG	ND	ND	ND	ND	++

Note: Cu is culicine, An is anopheline, T is tick, Ix is ixodid, Ar is argasid, Ph is phlebotomine, ND is no data, NG is no growth, Inc is inconsistent, and UNK is unknown.

[a] For classifications, refer to Reference 110.
[b] Cytopathic effect in clone C6/36.

pod cell cultures constitute a mixture in varied proportions of cells totally, partially, or completely nonpermissive to virus replication.

The quality of persistent infections depends upon the ratio of infected to noninfected cells[89] and may be defined as a "latent infection" (low ratio) or a "steady state of infection" (high ratio). The main features of persistence are summarized in Table 5.

A. Properties of Persistently Infected Cells

Slight CPE and most often formation of sincytia precedes the establishment of persistent infection,[90] but eventually unfused cells will participate in the total recovery of the culture as shown in cells infected with mosquito-borne flaviviruses.[91] Infected cells and noninfected cells do not exhibit distinctive features.

B. Viral and Nonviral Products of Persistence

1. Variants

Small plaque variants, thermosensitive mutants, and thermosensitive small plaque variants commonly occur in persistent infection of mosquito cells. Their appearance often follows the highest peak of virus production and they have been described for CHIK, SIN, and EEE.[92,93] The neutralization test does not detect any significant differences between the small plaque variants and the original wild strain. However the variants may exhibit a decreased pathogenicity for mice as reflected by lower or no mortality, or increased survival time. In some instances sensitive mutants did not revert after repeated passages in mammalian cells.[94]

2. Interfering Agents

The ability to resist challenge with a homologous or in rare instances with a heterologous virus is the result of by-products of viral persistence.

Defective interfering viral particles (deletion mutants) which can be new RNA species, pieces of RNA, or an incompletely formed viral genome will reduce cell damage and sometimes by competing with the production of normal virus will decrease its yield. Heterologous interference has rarely been reported,[95] whereas many have observed homologous and virus type specific interference. Thus *Aedes albopictus* cells persistently infected with the SIN virus will resist challenge with SIN but not with the closely related EEE virus or members of other taxons.[96-98]

Production of a substance with antiviral activity has been demonstrated in *A. albopictus* cells steadily infected with SIN.[99] This "interfering substance" was heat and pH labile (unlike interferon) and virus specific, with its effect nonreversible; it did not affect the adsorption of virus on cells and did not interfere with the normal replication of heterologous virus.

The resistance evolving from persistent infection of cell cultures either by small plaques or thermosensitive mutants, deletion mutants, or release of antiviral agents is with few exceptions not curable by treatment of the culture with antiserum to the original infecting strain.

The characteristics of persistent infections as observed in mosquito cell cultures in vitro may relate to natural infections of mosquitoes.[100] Coinfection of *A. albopictus* cells and clone C6/36 with two DEN virus types showed that: (1) cultures infected with any one type of four DEN viruses developed resistance with each of the other three types, (2) resistance as reflected by the proportion of cells containing antigen from superinfecting virus develop as early as 8 hr with only 40 to 50% of cells permissive to challenge virus and decreasing to only 0.1 to 1.0% after 20 hr, (3) induction of resistance is independent of cellular transcription since treatment of cells with Actynomycin D before and after infection with Dengue 1 did not prevent the occurrence of resistance to superinfection with DEN 3. Similar results were obtained when DEN 3 was the

Table 5
PARTICULAR ASPECTS OF ARBOVIRUS PERSISTENCE IN ARTHROPOD CELL CULTURES FROM SOME SELECTED STUDIES

Study	Cell culture	Observations	Ref.
CHIK	Aedes albopictus	No CPE, but syncitial formation	93
		Decrease of pathogenicity for mice directly related to transfer level of persistent culture	
CHIK	A. albopictus	No distinctive features between infected and noninjected cells	85
		Resistance to homologous superinfection	
		Persistence of greater duration with adapted virus strains	
SIN, SFV, EEE	A. aegypti	Small plaque mutants appearing after maximum viral peak	106
		Homologous interference only	
		When persistence occurs, there is a reduction of virus producing cells	
		Establishment of persistence is related to multiplicity of infection	
SIN	A. albopictus	Small plaque thermosensitive variants	96—98
		Impossibility to distinguish between variant and original strain	
		Variants remain thermosensitive even after passages in BHK21	
		Production of defective interfering particles	
MVE, KUN, D2, JE	A. albopictus	Establishment of persistence if preceded by syncitia	91
		Cell fusion is induced by inactivated virus	
		Small plaque mutants	
SIN	A. albopictus	Production of heat labile, pH labile, and virus-specific substance responsible for homologous interference	99
DEN	Toxorhynchites amboinensis (nonvector)	Increased ability to produce syncitia	107
		Reduction of neurovirulence for mice after one year, but no clear evidence of interference (DEN 1, 2, and 4)	
DEN	A. albopictus	Interference between closely related DEN 1 and 3 differs whether coinfection is simultaneous or not	100

primary infector and DEN 1 the superinfecting virus, (4) if coinfection with DEN 1 and 3 was simultaneous, the antigen to each virus type was detected in about 40 to 50% of the cells after 4 days and cells contained only one or the other type of antigen, but by day 20 both types of antigen could be found in a single cell. This observation is of epidemiologic importance since coinfection of mosquitoes with closely or distantly related arboviruses is possible. However it is not known if both viruses can be transferred simultaneously, if the ability to transmit is impaired by the production of viral recombinants or by replication of one virus at the expense of the other, and of what importance is the time which elapses between two infections. Concurrent overlapping periods of transmission of the *Alphavirus* WEE and the *Flavivirus* SLE (St. Louis encephalitis) by the same vector (*Culex tarsalis*) affects yearly fluctuations which cannot be entirely explained by the rapport of mosquitoes to reservoirs, but would satisfy the concept of viral heterotypic exclusion.

X. CONCLUSION

Despite their sophistication, cell culture systems are characterized by the mere maintenance of dedifferentiated cells which have lost at least part of their original in vivo functions and interactions and can only accommodate virus replication through a single metabolic pathway in a strictly controlled environment. Thus, the system is a poor mimicry of animal complexity. The ideal design of cell cultures tailored to the actual diversity of living systems and to their biologic fluctuations is probably far in the future and until then the actual methods can only serve specific purposes, such as isolation studies, antigenic mapping, and possibly production of vaccine strains.

The use of arthropod cell cultures for primary isolation of arboviruses holds some promise. However, its advantage over the conventional inoculation of mice or the more recent intrathoracic inoculation of mosquitoes (as yet not tested with many viruses) is not clear. Studies coordinated by various laboratories on selected viruses and according to a standard protocol are needed. The manufacture of more dipteran cell lines might only add redundancy to a catalogue of already well-tested and ascertained lines. On the other hand, the preferential growth of some arboviruses on either dipteran or tick cell lines, or both, calls for the greater variety of the latter.

Whether grown in vertebrate or mosquito cell cultures, arboviruses exhibit the same molecular weight, density, RNA structure, and polypeptide composition of the viral genome. Yet there are notable differences pertaining to each system of culture, the significance of which still has to be ascertained. It has thus been shown that some alpha- and flaviviruses grown in vertebrate cells lack a sialic acid component.[101,102] It has also been demonstrated that the qualitative make-up of the viral envelope which is derived from the host-cell membrane may exhibit changes modulated by the composition of this very membrane.[103] Such changes may also depend upon each cell type within the given cell-culture system. Comparative studies of SFV virus in *A. albopictus* and BHK21 and Kunjin in *A. albopictus* and Vero demonstrate such possible changes.[105] The antigenic drift or shift observed with arboviruses might in fact reflect a phenotypic expression acquired within a specific host (in vivo as well as in vitro) or through the alternation of hosts. The molecular analysis of virus replication in a variety of species of arthropod cells may illuminate some mechanisms of phenotypic changes and possibly nurture a tentative classification of host-related genetic traits which eventually will find their phylogenetic niche.

The natural ability of arthropod cell cultures to sustain persistent infection with arboviruses and to subsequently elaborate thermosensitive mutants makes them the ideal candidate for production of vaccine strains. However, the loss of virulence observed with some of these mutants is not necessarily accompanied by immunogenicity and their stability remains to be ascertained.

Our ability to isolate arboviruses and elucidate transmission mechanisms has greatly increased, but the natural history of arboviruses remains to be told and arthropod cell cultures may provide one of the most suitable methods to unravel the mystery of their evolution.

REFERENCES

1. Lloyd, W., Theiler, M., and Ricci, N. I., Modification of virulence of yellow fever virus in cultivation in tissues *in vitro, Trans. R. Soc. Trop. Med. Hyg.,* 29, 481, 1936.
2. Dulbecco, R., Production of plaques in monolayer tissue cultures by single particles of an animal virus, *Proc. Natl. Acad. Sci. U.S.A.,* 38, 747, 1952.
3. Dulbecco, R., Vogt, M., and Strickland, A. G. R., A study of the basic aspects of neutralization of two animal viruses, Western equine encephalitis virus and polyomyelitis virus, *Virology,* II, 162, 1956.
4. Frothingham, T. E., Tissue culture applied to the study of Sindbis virus, *J. Trop. Med. Hyg.,* 4, (Part 5), 863, 1955.
5. Bhatt, P. N. and Work, T. H., Tissue culture studies on arboviruses of the Japanese B-West Nile complex, *Proc. Soc. Exp. Biol. Med.,* 96, 213, 1957.
6. Buckley, S. M., Propagation, cytopathogenicity, and hemagglutination-hemadsorption of some arthropod-borne viruses in tissue culture, *Ann. N.Y. Acad. Sci.,* 81, 172, 1959.
7. Kissling, R. E., Growth of several arthropod-borne viruses in tissue culture, *Proc. Soc. Exp. Biol. Med.,* 96, 290, 1957.
8. Davies, N. C., Lloyd, W., and Frobisher, M., The transmission of neurotropic yellow fever virus by *Stegomya* mosquitoes, *J. Exp. Med.,* 56, 853, 1932.
9. Bates, M. and Garcia, R., Laboratory studies of the *Saimiri-Haemagogous* cycle of jungle yellow fever, *Am. J. Trop. Med.,* 25(3), 203, 1945.
10. Antunnes, P. C. and Whitman, L., Studies on the capacity of mosquitoes of the genus *Haemagogous* to transmit yellow fever, *Am. J. Trop. Med.,* 17, 825, 1937.
11. Davis, A. M. and Yoshpe-Purer, Y., Observation on the biology of West Nile virus with special reference to its behavior in the mosquito *Aedes aegypti, Ann. Trop. Med. Parasitol.,* 48, 46, 1954.
12. Hurlbutt, H. S. and Thomas, J. E., The experimental host range of the arthropod-borne animal viruses in arthropods, *Virology,* 12, 391, 1960.
13. Hurlbutt, H. S., Arthropod transmission of animal viruses, *Adv. Virus Res.,* 11, 277, 1965.
14. Lamotte, L. C., Japanese encephalitis virus in the organs of infected mosquitoes, *Am. J. Hyg.,* 72, 73, 1960.
15. Day, M. F. and Grace, T. D. E., Culture of insect tissues, *Annu. Rev. Entomol.,* 4, 38, 1959.
16. Jones, B. M. and Cunningham, I., Growth by cell division in insect tissue culture, *Exp. Cell Res.,* 23, 386, 1961.
17. Kurtti, T. J. and Munderloh, U. G., Mosquito cell culture, *Adv. Cell Cult.,* 3, 259, 1984.
18. Knudson, D. L. and Buckley, S. M., Invertebrate cell culture methods for the study of invertebrate-associated animal viruses, in *Methods in Virology,* Vol. 6, Maramorosch, K. and Koprowski, H., Eds., Academic Press, 1977, 323.
19. Jones, B. M., The cultivation of insect cells and tissues, *Biol. Rev.,* 37, 512, 1962.
20. Grace, T. D. C., Establishment of a line of mosquito (*Aedes aegypti* L.) cells grown in vitro, *Nature (London),* 211, 366, 1966.
21. Singh, K. R. P., Cell cultures derived from larvae of *Aedes albopictus* (Skuse) and *Aedes aegypti* (L.), *Curr. Sci.,* 36, 506, 1967.
22. Hink, F., the 1979 compilation of invertebrate cell lines and culture media, in *Invertebrate Systems In Vitro,* Kurstack, E., Maramorosch, K., and Dübendorfer, A., Eds., Elsevier/North Holland, New York, 1980, 553.
23. Pudney, M., Leake, C. J., and Buckley, S. M., Replication of arboviruses in arthropod *in vitro* systems, in *Invertebrate Cell Culture Applications,* Maramorosch, K. and Mitsuhashi, J., Eds., Academic Press, New York, 1982, 159.
24. Kurtti, T. J. and Munderloh, U. G., Tick cell culture: characteristics, growth requirements and applications to parasitology, in *Invertebrate Cell Culture Applications,* Maramorosch, K. and Mitsuhashi, J., Eds., Academic Press, New York, 1982, 195.

25. Grace, T. D. C., Establishment of four strains of cells for insect tissues grown in vitro, *Nature (London)*, 195, 788, 1962.
26. Cahoon, B. E., Hardy, J. L., and Reeves, W. C., Initiation and characterization of a diploid cell line from larval tissues of *Aedes dorsalis* (Meigen), *In Vitro*, 14, 255, 1978.
27. Varma, M. G. R., Pudney, M., and Leake, C. J., Cell lines from larvae of *Aedes (stegomya) malayensis* and *Aedesp seudoscuttellaris* (Theobald) and their infection with some arboviruses, *Trans. R. Soc. Trop. Med. Hyg.*, 68, 374, 1974.
28. Bhat, U. K. M. and Guru, P. Y., Establishment and characterization of larval cell lines, *Exp. Parasitol.*, 33, 105, 1973.
29. Sweet, B. H. and McHale, J. S., Characterization of cell lines derived from *Culiseta inornata* and *Aedes vexans* mosquitoes, *Exp. Cell Res.*, 61, 51, 1970.
30. Singh, K. R. P. and Bhat, U. K. M., Establishment of two mosquito cell lines from larval tissue of *Aedes w-albus*, *Experientia*, 27, 142, 1971.
31. Bhat, U. K. M. and Singh, K. R. P., Establishment of diploid cell line from larval tissues of *Aedes albopictus* (Bigot 1861), *Curr. Sci.*, 17, 388, 1970.
32. Marhoul, Z. and Pudney, M., A mosquito cell line (mos 55) from *Anopheles gambiae* larvae, *Trans. R. Soc. Trop. Med. Hyg.*, 66, 163, 1972.
33. Pudney, M. and Varma, M. G. R., *Anopheles stephensi* var. *mysorensis:* establishment of a larval cell line, *Exp. Parasitol.*, 29, 7, 1971.
34. Kitamura, S., Establishment of cell line from *Culex* mosquito, *Kobe J. Med. Sci.*, 16, 41, 1970.
35. Hsu, S. H., Mao, W. H., and Cross, J. H., Establishment of a line derived from ovarian tissue of *Culex quinquefasciatus* Say, *J. Med. Entomol.*, 7, 701, 1970.
36. Schneider, I., Establishment of cell lines from *Culex tritaeniorhynchus* and *Culex salinarius*, *Proc. 3rd Int. Colloq. on Inverteb. Tissue Cult.*, 1973, 121.
37. Chao, J. and Ball, G. H., The establishment of a *Culex pipiens* cell line and its adaptation to grow in three different insect tissue culture media. *In Vitro*, 8 (5), 406, 1973.
38. Hsu, S. H., Li, S. Y., and Cross, J. H., A cell line derived from ovarian tissues of *Culex tritaeniorhyncus*, *J. Med. Entomol.*, 9, 86, 1972.
39. Kuno, G., A continuous cell line of a non-hematophagous mosquito *Toxorhynchites amboinensis*, *In Vitro*, 16, 915, 1980.
40. Tesh, R. B., Establishment of two cell lines from mosquito *Toxorhynchites amboinensis* (Diptera: Culcidae) and their susceptibility to infection with arboviruses, *J. Med. Entomol.*, 17, 338, 1980.
41. Pudney, M. and Varma, M. G. R., Culture of embryonic cells from the tick *Boophilus micropolus*, *J. Med. Entomol.*, 10, 493, 1976.
42. Bhat, U. K. M. and Yunker, C. E., Establishment and characterization of a diploid cell line from the tick *Dermacentor parumapertus* Neumann, (Acarina: Ixodidae), *J. Parasitol.*, 63, 1092, 1977.
43. Yunker, C. E., Cory, J., and Meibos, H., Continuous cell lines from embryonic tissues of ticks, *In Vitro*, 17, 139, 1981.
44. Guru, P. Y., Dhanda, V., and Gupta, N. P., Cell cultures derived from the developing adults of three species of ticks by a simplified technique, *Indian J. Med. Res.*, 64, 1041, 1976.
45. Varma, M. G. R., Pudney, M., and Leake, C. G., The establishment of three cell lines from the tick *Rhipicephalus appendiculatus* (Acari:Ixodidae) and their infection with some arboviruses, *J. Med. Entomol.*, 11, 698, 1975.
46. Yunker, C. E., Arthropod tissue culture in the study of arboviruses and rickettsiae: a review, *Curr. Top. Microbiol. Immunol.*, 55, 113, 1971.
47. Buckley, S. M., Hayes, C. G., Maloney, J. M., Lipman, M., Aitken, T. H. G., and Casals, J., Arbovirus studies in invertebrate cell lines, in *Invertebrate Tissue Culture, Applications in Medicine, Biology and Agriculture*, Kurstak, E. and Maramorosch, K., Eds., Academic Press, New York, 1976, 3.
48. Stollar, V., Togaviruses in cultured arthropod cells, in *The Togaviruses*, Schlesinger, R. W., Ed., Academic Press, New York, 1980, 538.
49. Suitor, E. C., Growth of Japanese B encephalitis lines in Grace's continuous cell line of moth cells, *Biology*, 30, 143, 1966.
50. Converse, J. L. and Nagle, C., Jr., Multiplication of yellow fever virus in insect tissue cell cultures, *J. Virol.*, 1, 1096, 1967.
51. Trager, W., Multiplication of the virus of equine encephalitis in surviving mosquito tissues, *Am. J. Trop. Med. Hyg.*, 18, 387, 1938.
52. Peleg, F. L. and Trager, W., Cultivation of insect tissues in vitro and their application to the study of arthropod-borne viruses, *Am. J. Trop. Med. Hyg.*, 12, 820, 1963.
53. Peleg, J. and Trager, W., Growth of West Nile virus in insects tissues cultivated in vitro, *Ann. Epiphyt.*, 14, 211, 1963.
54. Rehacek, J. and Mayerova, M., Isolation of tick-borne encephalitis virus in tick haemocytes by the fluorescent antibody technique, *Acta Virol. Engl. Ed.*, 10, 374, 1976.

55. Cory, J. and Yunker, C. E., Rickettsial plaques in mosquito monolayers, *Acta Virol. Engl. Ed.*, 18, 512, 1974.
56. Yunker, C. E. and Cory, J., Plaque production by arboviruses in Singh's *Aedes albopictus* cells, *Appl. Microbiol.*, 19, 81, 1975.
57. Rehacek, J., The growth of arboviruses in mosquito cells in vitro, *Acta Virol. Engl. Ed.*, 12, 241, 1968.
58. Singh, K. R. P. and Paul, S. D., Isolation of dengue viruses in *Aedes albopictus* cell cultures, *Bull. W.H.O.*, 40, 982, 1962.
59. Chappell, W. A., Calisher, C. H., Toole, R. F., Maness, K. C., Sasso, D. R., and Henderson, B. E., Comparison of three methods used to isolate dengue virus type 2, *Appl. Microbiol.*, 22, 1100, 1971.
60. Varma, M. G. R., Pudney, M., Leake, C. G., and Peralta, P. H., Isolation in mosquito *Aedes psuedoscutellaris* of yellow fever virus strains from original field material, *Intervirology*, 6, 50, 1976.
61. Pavri, K. M. and Ghosh, S. N., Complement-fixation tests for simultaneous isolation and identification of dengue viruses, using tissue cultures, *Bull. W.H.O.*, 40, 984, 1969.
62. Race, M. W., Williams, M. C., and Agostini, C. F. M., Dengue in the Caribbean: virus isolation in a mosquito *Aedes pseudoscutellaris* cell line, *Trans. R. Soc. Trop. Med. Hyg.*, 73, 18, 1979.
63. Ajello, C. A., Evaluation of *Aedes albopictus* tissue culture for use in association with arbovirus isolation, *J. Med. Virol.*, 3, 301, 1979.
64. Henchal, E. A., McCown, J. M., Seguin, M. C., Gentry, M. K., and Brandt, W. E., Rapid identification of dengue virus isolates by using monoclonal anti-bodies in an indirect immunofluorescence assay, *Am. J. Trop. Med. Hyg.*, 31, 164, 1983.
65. Sweet, B. H. and Unthank, H. D., A comparative study of the viral susceptibility of monolayer and suspended mosquito cell lines, *Curr. Top. Microbiol. Immunol.*, 55, 150, 1971.
66. Tesh, R. B., A method for the isolation and identification of dengue viruses using mosquito cell cultures, *Am. J. Trop. Med. Hyg.*, 6, 1053, 1979.
67. De Mattos, I., Jozan Work, M., Work, T. H., Casals, J., and Buckley, S. M., Rapid diagnosis of St. Louis encephalitis virus in *Aedes albopictus* cell, in *Invertebrate Systems In Vitro*, Kurstak, E., Maramorosch, K., and Dübendorfer, A., Eds., Elsevier/North-Holland, New York, 1980, 347.
68. Lwoff, A. and Lwoff, M., Sur les facteurs du development viral et leur rôle dans l'évolution de l'infection, *Ann. Inst. Pasteur Paris*, 98, 174, 1960.
69. Brown, A., Differences in maximum temperature amongst selected group A arboviruses, *Virology*, 21, 362, 1963.
70. Johnson, J. W., Action of temperature growth of VEE and EEE in tissue culture of minced *Aedes aegypti* larvae, *Am. J. Trop. Med. Hyg.*, 18, 103, 1969.
71. Davey, M. W., Dennet, D. P., and Dalgarno, L., The growth of two togaviruses in cultured mosquito and vertebrate cells, *J. Gen. Virol.*, 20, 225, 1973.
72. Whitney, E. and Deibel, A., Growth studies of California encephalitis virus in two *Aedes* mosquito cell lines, *Curr. Top. Microbiol. Immunol.*, 55, 238, 1974.
73. Sarver, N. and Stollar, V., Sindbis virus induced cytopathic effects in clones of *Aedes albopictus* Singh cells, *Virology*, 80, 390, 1977.
74. Igarashi, A., Isolation of Singh's *Aedes albopictus* cell clone sensitive to dengue and Chikungunya viruses, *J. Gen. Virol.*, 40, 531, 1978.
75. Igarashi, A., Isolation of togaviruses by *Aedes albopictus* clone C6/36 cell culture, Communication at 11th Int. Congr. for Tropical Med. and Malaria, Calgary, Canada, September 1984.
76. Eaton, B. T., Transient enhanced synthesis of a cellular protein following infection of *Aedes albopictus* cells with Sindbis virus, *Virology*, 117, 307, 1982.
77. Reifel, F. and Kobler, H., Selective disappearance of two secreted host proteins in the course of Semliki Forest virus infection of *Aedes albopictus*, *J. Virol.*, 39, 31, 1981.
78. Ng, M. L. and Westaway, E. G., Phenotypic changes in the *Flavivirus* Kunjin after a single cycle of growth in an *Aedes albopictus* cell line, *J. Gen. Virol.*, 64, 1715, 1983.
79. Hirumi, H., Viral, microbial and extrinsic cell contamination of insect cell cultures, in *Invertebrate Tissue Culture: Research Applications*, Maramorosch, K., Ed., Academic Press, 1976, 233.
80. Hirumi, H., Hirumi, K., and Speyer, G., Further studies on the latent viruses isolated from Singh's *Aedes albopictus* cell line, in *Invertebrate Tissue Culture: Research Applications*, Maramorosch, K., Ed., Academic Press, 1976, 69.
81. Buckley, S. M., Susceptibility of the *Aedes albopictus* and *Aedes aegypti* cell lines to infection with arboviruses, *Proc. Soc. Exp. Biol. Med.*, 131, 625, 1969.
82. Singh, K. R. P., Growth of arboviruses in *Aedes albopictus* and *A. aegypti* cell lines, *Curr. Top. Microbiol. Immunol.*, 55, 127, 1971.
83. Singh, K. R. P., Goverdhan, M. K., and Bhat, U. K. M., Susceptibility of *Aedes w-albus* and *Anopheles stephensi* cell lines to infection with some arboviruses, *Indian J. Med. Res.*, 61, 2234, 1973.
84. Marhoul, Z., Susceptibility of *Anopheles gambiae* mosquito cell line to some arboviruses, *Acta Virol. Engl. Ed.*, 17, 507, 1973.

85. Buckley, S. M., Arboviruses and *Toxoplasma gondii* in diptera cell lines, in *Invertebrate Tissue Culture: Research Applications,* Maramorosch, K., Ed., Academic Press, New York, 1976, 201.

86. Rehacek, J., Cultivation of different viruses in tick tissue cultures, *Acta Virol. Engl. Ed.,* 9, 332, 1965.

87. Pudney, M., Varma, M. G. R., and Leake, C. J., The growth of some arboviruses in tick cell lines, in *Tick-Borne Diseases and Their Vectors,* Wilde, J. K. H., Ed., Lewis Reprints, Tonbridge, 1978, 490.

88. Leake, C. J., Pudney, M., and Varma, M. G. R., Studies on arboviruses in established tick cell lines, in *Invertebrate Systems In Vitro,* Kurstak, E., Maramorosch, K., and Dübendorfer, A., Eds., Elsevier/North Holland, New York, 1980, 327.

89. Fenner, F., McAuslan, B. R., Mims, C. A., Sambrook, J., and White, D. O., *Biology of Animal Virus,* Academic Press, New York, 1974, 452.

90. Banerjee, K. and Singh, K. R. P., Establishment of carrier cultures of *Aedes albopictus* cell lines infected with arboviruses, *Indian J. Med. Res.,* 56, 812, 1968.

91. Ng, M. L. and Westaway, E. G., Establishment of persistent infections by flaviviruses in *Aedes albopictus* cells, in *Invertebrate Systems In Vitro,* Kurstak, E., Maramorosch, K., Dübendorfer, A., Eds., Elsevier/North Holland, New York, 1980, 389.

92. Buckley, S. M., Singh's *Aedes albopictus* cell cultures as helper cells for the adaptation of Obodhiany and Kotonkan viruses of the rabies sero-group to some vertebrate cell cultures, *Appl. Microbiol.,* 25, 695, 1973.

93. Banerjee, K. and Singh, K. R. P., Loss of mouse virulence in Chikungunya virus from the carrier culture of *Aedes albopictus* cell line, *Indian J. Med. Res.,* 57, 1003, 1968.

94. Shenk, T. E., Koshelnyk, K. A., and Stollar, V., Temperature sensitive virus form *Aedes albopictus* cells chronically infected with Sindbis virus, *J. Virol.,* 13, 439, 1984.

95. Eaton, B. T., Heterologous interference in *Aedes albopictus* cells infected with alphaviruses, *J. Virol.,* 30, 45, 1979.

96. Shenk, T. E. and Stollar, V., Defective interfering particles of Sindbis virus, *Virology,* (Part I), 53, 162, 1973a.

97. Shenk, T. E. and Stollar, V., Defective interfering particles of Sindbis virus, *Virology,* (Part II), 55, 531, 1973b.

98. Stollar, V. and Shenk, T. E., Homologous viral interference in *Aedes albopictus* cell cultures chronically infected with Sindbis virus, *J. Virol.,* 11, 529, 1973.

99. Riedel, B. and Brown, D., Novel antiviral activity found in the media of Sindbis virus persistently infected *Aedes albopictus* cell cultures, *J. Virol,* 29, 51, 1979.

100. Dittmar, D., Castro, H., and Haines, H., Demonstration of interference between dengue virus types in cultured mosquito cells using monoclonal antibody probes, *J. Virol.,* 59, 273, 1982.

101. Luukkonen, B., Kaariainen, L., and Renkonen, O., Phospholipids of Semliki Forest virus grown in cultured mosquito cells, *Biochim. Biophys. Act.,* 450, 109, 1976.

102. Stollar, V., Togaviruses in cultured arthropod cells, in *The Togaviruses,* Schlesinger, R. W., Ed., Academic Press, New York, 1980, 583.

103. Westaway, E. G., Replication of Flaviviruses, in *The Togaviruses,* Schlesinger, R. W., Ed., Academic Press, New York, 1980, 531.

104. Morita, K., Hori, H., and Igarashi, A., Fingerprint analysis on Getah and Japanese encephalitis viruses, Communication at 11th Int. Congress for Tropical Med. and Malaria, Calgary, Canada, September 1984.

105. Luukkonen, A., Von Bonsdorff, C. H., and Renkonen, O., Characterization of Semliki Forest virus grown in mosquito cells. Comparison with the virus from hamster cells, *Virology,* 78, 331, 1977.

106. Peleg, J. and Stollar, V., Homologous interference in *Aedes aegypti* cultures infected with Sinbis virus, *Archiv. Gesamte Virusforsch.,* 45(4), 309, 1974.

107. Kuno, G., Persistent infection of a non-vector mosquito cell line (TRA-171) with dengue viruses, *Intervirology,* 18, 44, 1982.

108. Work, T. H. and Jozan, M., Unpublished information.

109. Schneider, I., Unpublished information, 1974.

110. Matthews, R. E. F., *Intervirology,* 7, 1, 1982.

Cultivation and Characterization of Arthropod Cells as Substrates for Virus Growth

Chapter 2

PREPARATION AND MAINTENANCE OF ARTHROPOD CELL CULTURES: DIPTERA, WITH EMPHASIS ON MOSQUITOES

I. Schneider

TABLE OF CONTENTS

I. INTRODUCTION

The first purported dipteran cell line to be established was from the mosquito *Aedes aegypti* L. by Grace of Australia.[1] The primary cultures consisted of the alimentary canals, minus the peritrophic membranes, and bodies from axenically reared fourth instar larvae on the verge of pupation. In a complex medium formulated in part on an analysis of hemolymph from the silkworm, *Bombyx mori*[2] and supplemented with hemolymph from diapausing pupae of yet another moth, *Antheraea eucalypti,* the cells from the explants grew in sufficient number to permit successful subculturing after 4 weeks in vitro. The validity of this line has been questioned as both the morphology and chromosome complement of the cells resembled that of the first lepidopteran cell lines established a few years earlier, also in the laboratory of Grace.[3] Whether of a bona fide *Aedes aegypti* line or not, this report served to rekindle interest in the possibility of obtaining continuous lines from dipteran species[4,5] and led shortly thereafter to the demonstration by Singh of Poona, India, that cell lines could be established using a much simplified medium and with fetal bovine serum (FBS) rather than homologous or heterologous hemolymph as the medium supplement.[6] The techniques of Grace and Singh, particularly the latter, have since been standardized albeit with many variations and used for the initiation and establishment of virtually all dipteran cell lines now in existence. These currently number in excess of 300 with almost 70 of the lines derived from mosquito species. Most of the remainder are from *Drosophila.*

II. PRIMARY CULTURES

A. Source of Explants

The great majority of primary cultures from dipterans have been initiated with neonate larvae that have been minced into two or three fragments and then subjected to enzymatic treatment.[7-9] Trypsin has been the enzyme most often used at a concentration of 0.1 to 0.25% in Rinaldini's salt solution (RSS)[10] (NaCl — 800 mg; KCl — 20 mg: $NaH_2PO_4 \cdot H_2O$ — 5 mg; $NaHCO_3$ — 100 mg; sodium citrate — 67.6 mg; glucose — 100 mg; distilled H_2O to 100 mℓ) or in other phosphate buffered Ca- and Mg-free saline.[8] However, trypsin/EDTA (0.5/0.2%), collagenase (0.2%), hyaluronidase (0.2 to 0.25%), pronase (0.1%), and chitinase (0.25%) in RSS have all proved to be equal or similar in effectiveness. Usually 100 to 200 larvae/1.5 mℓ medium are required to insure a successful primary culture, although with some mosquito genera, such as *Toxorhynchites,* 10 to 15 may be sufficient. The larvae are obtained in sterile condition by allowing them to hatch after surface sterilization of the eggs (see next section). Thus there is virtually no possibility of contamination other than that which may be present within the cells of the primary donors. Indigenous viruses or virus-like particles are, however, by no means rare in cells of immature and adult insects examined by electron microscopy.[11,12]

The neonate larvae are usually minced with fine scissors or a small, sharp blade; 1 mℓ syringes fitted with No. 27 gauge needles are very suitable for this purpose. The larvae may be partially immobilized by reducing the amount of fluid (either saline or medium lacking FBS) surrounding them in a depression slide or by chilling them. Once minced, the fragments are either centrifuged at low speed (300 rpm for 5 min) or transferred to a small diameter (8 to 10 mm) glass or plastic tube with a retention screen at the bottom. With either method the supernatant or excess fluid can be withdrawn and the enzyme solution added. The larval fragments are left in the solution for 10 to 60 min at temperatures ranging from 25 to 37°C. Following treatment the fragments are first washed with serum or placed directly in complete medium, e.g., medium plus FBS.

A number of cell lines have also been initiated from crushed embryos or from larval fragments that have not been subjected to enzyme treatment,[13-16] but this is inadvisable for *Aedes* species and probably for anophelines as well (unpublished observations) in that cell attachment, migration, and growth tend to be less extensive for these species and the interval from setting up the primary to the first successful subculture may take as long as 4 to 6 months.[13] The use of larvae past the first instar is usually precluded by the appearance in the hemolymph of the enzyme tyrosinase (= phenol oxidase) which produces melanin via intermediary, possibly toxic quinones.[2] In such cultures the presence of the tyrosinase is evidenced by the medium turning dark brown or black after 24 to 48 hr of incubation. Bands of melanin invariably form at the cut ends of the larval fragments effectively precluding both cell migration and growth.[17] Whether inhibitors such as glutathione or ascorbic acid would suppress this enzyme in a culture system has not been adequately addressed.

Additionally, intact or minced adult mosquito ovaries have infrequently served as a source of primary explants with five or more pairs placed in 3 mℓ of medium.[18-21] The females are immersed for 2 min in a 0.25% solution of $HgCl_2$ in 70% ethanol, washed in distilled water, and the ovaries removed in a suitable medium. No enzymatic treatment is applied. Here also initial cell migration and growth during the first month of culture often exceeds that from larval explants but may lag thereafter with subculturing, requiring as little as 3 weeks[18] to as much as 7 months.[21] One great disadvantage of using organ explants from adults or late larvae is the ever present possibility of contamination from inadvertently damaging the digestive tract during dissection. The use of axenically reared insects remains a possibility, but past attempts to rear mosquitoes under such conditions have been met with limited success.[22-25] However, the recent and novel method of Munderloh et al.[16] in axenically rearing *Anopheles stephensi* and *Toxorhynchites amboinensis* on a continuous *T. amboinensis* cell line, if applicable to other genera as well, should circumvent this problem.

B. Sterilization Techniques

Surface sterilization of the eggs should be carried out within 24 to 36 hr of hatching, as younger embryos may not withstand the procedure. If the age of the eggs from mosquito species are unknown, they may be sterilized any time after the appearance of red eye spots usually visible through the chorion. The initial step in sterilization is designed to remove the very thin exochorion and for this purpose the eggs are immersed in a 1:1 solution of water and household bleach (equivalent to a 2.5% solution of sodium hypochlorite) for 1 or 2 min at room temperature. A longer interval may harm the embryo and/or render the eggs so sticky that they are virtually impossible to handle further. The eggs are then surface sterilized using any number of agents for 2 to 20 min. Among the more common are 70% ethanol, 0.05% $HgCl_2$ in 70% ethanol, White's solution ($HgCl_2$ — 25 mg; HCl — 0.125 mℓ; NaCl — 650 mg; 70% ethanol — 25 mℓ, and distilled H_2O — 75 mℓ)[26] and 0.1% benzalkonium chloride (= Roccal®). After several rinses the eggs are transferred to a petri dish containing moistened, sterile filter paper and the larvae are allowed to hatch. Embryos from those species requiring "conditioning" before hatching are first freed of as much debris as possible, placed on several layers of filter paper, and allowed to dry for a week or longer at room temperature. Immersion in 0.1% benzalkonium chloride followed by rinses in 0.01 *M* phosphate buffer is advisable for sterilizing these eggs, as premature hatching in any of the ethanol solutions results in immediate death. To ensure virtually synchronous hatching of the larvae, the eggs are placed in a desiccator and subjected to reduced pressure under vacuum.

Table 1
INVERTEBRATE MEDIA MOST WIDELY USED FOR MOSQUITO
CELL CULTURE

Components[a]	M&M	VP12	M-Hsu	M-Kitamura
$NaH_2PO_4 \cdot H_2O$	20	48.6	—	—
$MgCl_2 \cdot 6H_2O$	10	110	—	—
$MgSO_4 \cdot 7H_2O$	—	120	14	—
KCl	20	55	28	50
KH_2PO_4	—	—	8.75	10
$CaCl_2 \cdot H_2O$	20	40	16.2	10
NaCl	700	390	157.5	650
$NaHCO_3$	12	50	30	10
D-glucose	400	200	56	400
Sucrose	—	—	210	—
Lactalbumin hydrolysate	650	500	865	650
Yeastolate	500	25	320	500
Bactopeptone	—	—	175	—
Bovine albumin powder (Fraction V)	—	100	—	—
Choline chloride	—	25	—	—
Inositol	—	40	—	—
1-Malic acid	—	—	21	—
α-Ketoglutaric acid	—	—	14	—
Succinic acid	—	—	2.1	—
Fumaric acid	—	—	2.1	—
Glutamine (5%)	—	0.6 ml	—	—
Basal medium (Eagle) vitamin mixture (100×)	—	2 ml	—	—
Medium 199 (10×)	—	—	5 ml	—
Whole egg ultrafiltrate	—	—	1 ml	—
Fetal bovine serum	20%	10%	10%	20%

Note: M&M is Mitsuhashi and Maramorosch;[26] VP12 is Varma and Pudney;[31] M-Hsu is Modified Hsu;[19] and M-Kitamura is Modified Kitamura.[21]

[a]   In mg/100ml unless otherwise stated.

C. Media Formulations

Culture media have been patterned after the hemolymph composition of a particular insect, derived empirically or borrowed from vertebrate cell culture.[8,9,27,28] Among the latter, Leibovitz's L-15 has been the most popular, with Medium 199 and Eagle's somewhat less so.[16,29,30] Undoubtedly the most widely used medium for mosquito cells is that designed by Mitsuhashi and Maramorosch for cultures of leafhopper embryonic, nymphal, and adult stages.[26] Singh[6] first demonstrated that this Mitsuhashi and Maramorsch medium adequately supported the growth of *Aedes* cells and it has proven equally suitable for some *Culex* and *Toxorhynchites* species as well (Tables 1 and 2). In addition to its adequacy as a medium, its popularity also stems from the modest number of components, e.g., individual amino acids are replaced by lactalbumin hydrolysate and the B vitamins by a yeast extract, and from its ease of preparation. Other media used with some frequency are those of Hsu,[18-20] Kitamura,[21] and Varma and Pudney.[31]

Most media are supplemented with 10 to 20% heat inactivated (56°C for 30 min) FBS. The earlier practice of using heat-treated hemolymph, egg albumin, insect extracts, or chick embryo extract has been largely discontinued. Frequently the percentage of FBS is reduced as the number of passages increases. In a number of instances, well-established lines have been adapted to grow in serum-free media by gradually reducing the serum concentration over a period of weeks or months, but not without

Table 2
CONTINUOUS MOSQUITO CELL LINES

Genus and species	Source of primary explant	No. of lines	Medium utilized	Ref.
Aedes aegypti	Fourth instar larvae	1	Grace/5% hemolymph	1
	Neonate larvae	2	M&M/20%FBS	6
	Neonate larvae	3	M&M, M&M/M-K or M&M/VP12 with 20%FBS	31
	Embryos	2	K/10%FBS/10%CM/1%CEE	13,43
A. albopictus	Neonate larvae	3	M&M/20%FBS	6
	Adult ovaries	1	M&M,Hsu or M-Hsu with 10%FBS	20
	Neonate larvae	1	M&M/20%FBS	32
A. dorsalis	Neonate larvae	1	M-K or M&M with 20%FBS	44
A. malayensis	Neonate larvae	1	M&M, M&M/M-K or L-15 with 15—20%FBS	45
A. novalbopictus	Neonate larvae	2	M&M/20%FBS	46
A. pseudoscuttellaris	Neonate larvae	2	M&M, M&M/M-K or L-15 with 15—20%FBS	45
A. vittatus	Neonate larvae	1	M-Grace/10%FBS	37
A. w-albus	Neonate larvae	2	M&M/20%FBS	47
Anopheles gambiae	Neonate larvae	1	M-K/VP12 with 15%FBS	48
An. stephensi	Neonate larvae	3	M-Grace/15%FBS	49
	Neonate larvae	4	M-K/VP12 with 15%FBS	50
Culex bitaeniorhynchus	Embryos	1	M&M/20%FBS	15
C. pipiens	Adult ovaries	1	K/10%FBS	21
	Embryos	1	M&M, Hsu or Schneider with 10%FBS	51
C. quinquefasciatus	Adult ovaries	2	Hsu/10%FBS	18
C. salinarius	Neonate larvae	1	M-Hsu/15%FBS	52
C. tarsalis	Embryos	1	M&M, Hsu or Schneider with 10%FBS	53
C. tritaeniorhynchus	Adult ovaries	1	M-Hsu/10%FBS	19
	Neonate larvae	1	M-Hsu/15%FBS	52
Haemogogus equinus	Neonate larvae	1	L-15/MEM/15%FBS	54
Toxorhynchites amboinensis	Embryos, neonate larvae	2	M&M/VP12/15%FBS	14
	Neonate larvae	23	M&M/VP12/10—15%FBS	55,56
	Embryos	1	L-15/tryptose phosphate broth/20%FBS	16

Note: CEE is chick embryo extract; CM is conditioned medium; MEM is minimal essential medium; M&M is Mitsuhashi and Maramorosch[26]; K is Kitamura[21]; L-15 is Leibovitz's L-15; the prefix M- is modified; VP12 is Varma and Pudney[31]; and FBS is fetal bovine serum.

some consequence to growth rates, seeding density requirements, ploidy, and phenotype.[30,32] Additionally at least one chemically defined medium has been used to support the growth of two continuous lines from *Aedes aegypti* and *Anopheles gambiae* for a minimum of five passages, apparently without the need for any adaptation period.[33]

Although pH values of dipteran hemolymph range between 6.3 to 7.7, and are known to fluctuate during different stages of the life cycle,[28] the pH of most culture media falls between 6.8 to 7.0. Whether the cells actually require such a limited range is unknown but it seems unlikely. Hsu, for instance, reported that *Culex tritaeniorhynchus* cells were capable of growing equally well in media having a pH range from 6.3 to 7.6.[19] Likewise the osmotic pressure of insect hemolymph varies from stage to stage of a single individual, being dependent on both internal and external factors. Hence, the osmolality of a medium may not be that critical for cultured cells and indeed some cell lines have been grown in both vertebrate and invertebrate media (290 vs. up to 400 mOsm/ℓ) with little adverse effect.[28,34]

D. Other Parameters

Temperatures between 26 and 30°C are normally used although cells have been successfully grown at extremes as low as 15°C[35] and as high as 35°C.[13] The growth rate is invariably much slower at the lower temperatures and often at the higher temperatures as well. In a few instances the cultures have been gassed with CO_2 in air, but the great majority of dipteran cell lines are grown in an ambient atmosphere. A medium depth of 3 to 4 mm appears to be optimal. Dipteran cells in primary culture tend to attach more tenaciously to glass than to plastic and hence glass vessels are recommended for that reason. Once subculturing has been successful, the usual practice is to grow the cells in T-30 or T-75 plastic, disposable flasks.

The routine use of antibiotics and antimycotics (penicillin — 100 units/mℓ; streptomycin — 100 μg/mℓ; gentamycin sulfate — 50 to 100 μg/mℓ; ampotericin B — 2 to 5 μg/mℓ) is to be avoided, as they may mask cryptic contaminants. Unless the cells are deemed irreplaceable, it is best to discard any contaminated culture rather than attempt to rid it of the infection.

III. EVOLUTION OF CELL LINES

A. Initial Growth Patterns and Cell Types

One of the most distinctive characteristics of primary cultures from most dipteran species, first noted by Singh in *Aedes aegypti* cultures,[6] is the growth of cellular vesicles issuing from the cut ends of the larval fragments. These single-layered vesicles often appear within a few days of initiating the primaries and in time may consist of hundreds or many thousands of cells. The fragment ends may support a single vesicle or considerable numbers, and additional vesicles usually form if the initial ones spontaneously detach or are excised. This type of growth may continue for weeks or months and in a few cases may continue well after the original larval fragments are eliminated through repeated subculturing, e.g., the vesicles are capable of propagating themselves by a type of budding.[16,36] The appearance of these vesicles is not a function of osmotic pressure, as they arise in iso-, hypo-, and hypertonic media. Their development, however, is undoubtedly encouraged by the enzymatic treatment which is designed to loosen the matrix between the cells to encourage migration rather than provide an inoculum of single cells or small groups of cells. Usually, however, the bottom of the culture vessel becomes dotted with cell colonies from groups of stray cells, from detached vesicles that break open to form cell sheets, and from larval fragments that attach directly to the glass or plastic substrate. Under favorable conditions these colonies increase by cell division and cover much of the growing area within 5 to 20 days. Sometimes, however, cell migration predominates over division and the culture enters a quiescent state which may last for many months before resumption of growth. Whether the latter is indicative of transformation as seen in vertebrate cell culture has been the subject of much speculation but little consensus.

Attachment of intact or minced ovaries to the glass substrate is a prerequisite for successful culture. Pulsation movements are common and may last for up to 7 weeks after initiating the culture. Vesicle formation has not been reported.

With the exception of imaginal disc cell lines, the identity of the cells which survive in culture is unknown. Cells which can be identified on the basis of morphological or physical characteristics such as nerve, muscle, trachea, and macrophages usually disappear by the 2nd or 3rd week of culture. The cell types remaining are classified solely by their growth pattern, e.g., epithelial, fibroblast-like, or hemocyte-like.

It is advisable to partially renew the medium in primary cultures by a third or a half every 7 to 10 days. As cell multiplication becomes more evident the interval may need to be shortened.

B. Subculturing

Subculturing attempts are usually made when the cells in a primary culture cover approximately two thirds of the bottom of the culture vessel. Cultures in which the cells can be dislodged (especially by pipetting) and that reattach within 30 min are good candidates for subculturing.

Reports on the interval between initiating a primary culture and the first successful subculture range from less than 2 weeks[14,15,36,37] to as long as 7 months.[21] Whether this is a measure of relative tolerance to in vitro culture per se or simply a reflection of inadequate media and/or other culture conditions is unknown. It is obvious, however, that with the present state of the art, the cells of certain genera are far more amenable to culture than others. Of the mosquito genera represented by more than one continuous line, the order from the easiest established to the most difficult is: *Toxorhynchites > Culex > Aedes > Anopheles.*

IV. CONTINUOUS CELL LINES

A. Genera and Species Represented

A listing of the continuous mosquito lines established thus far is given in Table 2 along with details on the source of the primary explants, number of lines established, media utilized, and the pertinent reference. Until recently most insect cell lines were designated only by the genus and species from which they arose and confusion inevitably arose as multiple lines from a single species were established. This situation has now been somewhat mitigated by the increasingly common practice of using a hyphenated-three-part designation that includes two to five letters identifying the laboratory where initiated, upper and lower case letters for genus and species, respectively, and a number or number-letter combination identifying the cell line.[28]

Most of the cell lines listed can be subcultured simply by dislodging the cells by pipetting and transferring an aliquot to fresh medium. Once in continuous culture it is not unusual to split the cells 1:10 or even 1:100. Population doubling times for the different lines are contingent upon temperatures, with most ranging from 16 to 24 hr at 28 to 30°C and from 24 to 30 hr at 24 to 28°C.[9]

B. Chromosome Analysis

A simple and rapid method for monitoring cell populations is by chromosomal analysis. This is especially true for mosquito cells, as all species have a diploid number of six and the chromosomes have a distinctive morphology. Although rarely adequate to distinguish between mosquito genera, the technique has been used to detect cell lines labeled as mosquito that were either lepidopteran in origin or had been contaminated with lepidopteran cells.[1,38] Given adequate culture conditions, dipteran cell lines tend to be genetically stable even after many passages in vitro. Kurtti and Munderloh[9] noted that 27 of 33 mosquito cell lines had a modal chromosome number of six, two were tetraploid, and four either were not tested or were tested with equivocal results.

The chromosomes are prepared by incubating the cells in the midlog phase of growth with Colcemid® (0.06 to 0.12 μg/ℓ) for 4 to 18 hr at 25°C. The cells are then flushed from the bottom of the flask, transferred to one or more 15 mℓ centrifuge tubes, and the tubes placed in a 30 to 32°C water bath. Equal volumes of prewarmed distilled water are added at times 0, 10, and 20 min for a total dilution of 1:8. Freshly prepared acetic acid:methanol fixative (1:3) is slowly added using 8 drops of fixative per 10 mℓ

diluted medium. The cells are centrifuged at 400 rpm for 5min, the supernatant discarded, and 0.1 mℓ fixative added to the pellet; 20 min later the cells are resuspended in additional fixative and again spun at 400 rpm. Most of the supernatant is withdrawn and aliquots of the remaining cell suspension dropped onto precleaned slides that have been dipped in distilled water. After air drying for at least 2 hr, the cells are stained with a freshly prepared solution of Giemsa.

C. Cloning Methods

The heterogeneity of cell types in continuous lines (especially those derived from several hundred individuals) may make it desirable to clone the cells for any number of purposes, e.g., for metabolic studies, to isolate biochemical variants, or for greater susceptibility to viral infection than the parental line. The much exploited C6/36 clone isolated by Igarashi[39] from Singh's *Aedes albopictus* line attests to the usefulness of the technique.

Cloning by limiting dilution is probably the most widely used method.[40] The cells are diluted to a concentration of one cell or less per mℓ medium and 0.2 to 0.3 mℓ aliquots transferred to each well of a plastic 96 well plate. The presence of a single cell per well may be checked using an inverted microscope. Viable clones are transferred to larger culture vessels as soon as cell density permits, usually in 2 to 3 weeks. Equally effective but much more laborious was Igarashi's procedure of seeding a dilute suspension of cells into petri dishes and transferring the resulting individual colonies with small pieces of trypsin/EDTA soaked filter paper.[39] Colony formation in agar medium is yet another alternative. A base layer of 1% agar in heart infusion broth and medium is allowed to solidify in a small petri dish before a dilute (0.3%) cell suspension in special Agar-Noble is placed on top. The plates are sealed to prevent moisture loss, and in 10 to 14 days the resulting colonies are removed with a Pasteur pipet and placed in the appropriate medium.[41,42] Plating efficiencies of the cells tend to be 1% or less.[28]

D. Cryopreservation

Dipteran cell lines may be cryopreserved using 10% glycerol, 10% dimethylsulfoxide (DMSO), or 5% of each in complete medium. With the advent of polypropylene cryotubes, the use of potentially hazardous glass ampules can be avoided. Cell concentrations of 1×10^6 or 10^7/mℓ are recommended. The tubes may simply be wrapped in cotton and placed in a $-80°C$ freezer. After 24 hr it is advisable to transfer the tubes to liquid nitrogen. The cells remain viable for up to a year at $-80°C$ and for well over 5 years at $-190°C$. The cells are thawed by placing the tubes in a 30 to 32°C water bath and gently shaking. Once thawed, the cells are appropriately diluted in the medium. If glycerol is used as the cryopreservative, it is advisable to replace a portion or all of the medium 24 hr later. It is not uncommon to have recovery rates of over 90%.

REFERENCES

1. Grace, T. D. C., Establishment of a line of mosquito (*Aedes aegypti* L.) cells grown *in vitro*, *Nature*, 211, 266, 1966.
2. Wyatt, S. S., Culture *in vitro* of tissue from the silkworm, *Bombyx mori* L., *J. Gen. Physiol.*, 39, 841, 1956.
3. Grace, T. D. C., Establishment of four strains of cells from insect tissues grown *in vitro*, *Nature*, 195, 788, 1962.
4. Day, M. F. and Grace, T. D. C., Culture of insect tissues, in *Annual Review of Entomology*, Vol. 4, Steinhaus, E. A., Ed., Annual Reviews, Stanford University, Palo Alto, Calif., 1959, 17.

5. Brooks, M. A. and Kurtti, T. J., Insect cell and tissue culture, in *Annual Review of Entomology*, Vol. 16, Annual Reviews, Stanford University, Palo Alto, Calif., 1971, 27.
6. Singh, K. R. P., Cell cultures derived from larvae of *Aedes albopictus*(Skuse) and *Aedes aegypti* (L.), *Curr. Sci.*, 361, 506, 1971.
7. Schneider, I., Initiation of primary cultures from dipteran species by enzymatic dissociation, in *TCA Manual*, Vol. 5, Tissue Culture Association, Gaithersburg, Md., 1979, 987.
8. Varma, M. G. R., Pudney, M., and Leake, C. J., Methods in mosquito cell culture, in *Practical Tissue Culture Applications*, Maramorosch, K. and Hirumi, H., Eds., Academic Press, New York, 1979, 331.
9. Kurtti, T. J. and Munderloh, U. G., Mosquito cell culture in *Advances in Cell Culture*, Vol. 3, Maramorosch, K., Ed., Academic Press, New York, 1984, 259.
10. Rinaldini, L. M., A quantitative method for growing animal cells *in vitro, Nature*, 173, 1134, 1954.
11. Hirumi, H., Viral, microbial and extrinsic contamination of insect cell cultures, in *Invertebrate Tissue Culture, Research Applications*, Maramorosch, K., Ed., Academic Press, New York, 1976, 233.
12. Hirumi, H., Hirumi, K., Speyer, G., Yunker, C. E., Thomas, L. A., Cory, J., and Sweet, B. H., Viral contamination of a mosquito cell line, *Aedes albopictus,* associated with syncytium formation, *In Vitro*, 12, 83, 1976.
13. Peleg, J., Growth of arboviruses in monolayers from subcultured mosquito embryo cells, *Virology*, 35, 617, 1968.
14. Tesh, R. B., Establishment of two cell lines from the mosquito *Toxorhynchites amboinensis*(Diptera: Culicidae) and their susceptibility to infection with arboviruses, *J. Med. Entomol.*, 17, 338, 1980.
15. Pant, U. and Dhanda, V., Establishment of a cell line from *Culex bitaeniorhynchus, J. Tissue Cult. Methods.*, 6, 61, 1980.
16. Munderloh, U. G., Kurtti, T. J., and Maramorosch, K., *Anopheles stephensi* and *Toxorhynchites amboinensis:* aseptic rearing of mosquito larvae on cultured cells, *J. Parasitol.*, 68, 1085, 1982.
17. Schneider, I. and Vanderberg, J. P., Culture of the invertebrate stages of plasmodia and the culture of mosquito tissues, in *Malaria*, Vol. 2, Kreier, J. P., Ed., Academic Press, New York, 1980, 235.
18. Hsu, S. H., Mao, W. H., and Cross, J. H., Establishment of a line of cells derived from ovarian tissue of *Culex quinquefasciatus* Say, *J. Med. Entomol.*, 7, 703, 1970.
19. Hsu, S. H., Li, S. Y., and Cross, J. H., A cell line derived from ovarian tissue of *Culex tritaeniorhynchus* Summorosus Dyar, *J. Med. Entomol.*, 9, 86, 1972.
20. Hsu, S. H., Huang, M. H., Wang, W. J., and Lin, S. N., Growth of some togaviruses in *Aedes albopictus* (Diptera: Culicidae) cell cultures, *J. Med. Entomol.*, 14, 581, 1978.
21. Kitamura, S., Establishment of cell line from *Culex* mosquito, *Kobe J. Med. Sci.*, 16, 41, 1970.
22. Akov, S., Qualitative and quantitative study of nutritional requirements of *Aedes aegypti* L. larvae, *J. Insect Physiol.*, 8, 319, 1962.
23. Rosalis-Ronquillo, M. C., Simons, R. W., and Silverman, P. H., Aseptic rearing of *Anopheles stephensi* (Diptera: Culicidae), *Ann. Entomol. Soc. Am.*, 66, 949, 1973.
24. Dadd, R. H. and Kleinjan, J. E., Chemically defined dietary media for larvae of the mosquito *Culex pipiens* (Diptera: Culicidae): effects of colloid texturizers, *J. Med. Entomol.*, 13, 285, 1976.
25. Sneller, V.-P. and Dadd, R. H., *Brugia pahangi:* development in *Aedes aegypti* reared axenically on a defined synthetic diet, *Exp. Parasitol.*, 51, 169, 1981.
26. Mitsuhashi, M. and Maramorosch, K., Leafhopper tissue culture: embryonic, nymphal and imaginal tissues from aseptic insects, *Contrib. Boyce Thompson Inst.*, 22, 435, 1964.
27. Mitsuhashi, J., Media for insect cell cultures, in *Advances in Cell Culture*, Vol. 2, Maramorosch, K., Ed., Academic Press, New York, 1982, 133.
28. Vaughn, J. L., Insect tissue culture: techniques and development, in *Techniques in the Life Sciences*, Kurstak, E., Ed., Elsevier, Amsterdam, 1985, chap. C108, 1.
29. Kuno, G., A continuous cell line of a nonhematophagous mosquito, *Toxorhynchites amboinensis, In Vitro*, 16, 915, 1980.
30. Kuno, G., Cultivation of mosquito cell lines in serum-free media and their effects on dengue virus replication, *In Vitro*, 19, 707, 1983.
31. Varma, M. G. R. and Pudney, M., The growth and serial passage of cell lines from *Aedes aegypti* (L.) larvae in different media, *J. Med. Entomol.*, 6, 432, 1969.
32. Mitsuhashi, J., A new continuous cell line from larvae of the mosquito *Aedes albopictus* (Diptera: Culicidae), *Biomed. Res.*, 2, 599, 1981.
33. Wilkie, G. E. I., Stockdale, H., and Pirt, S. V., Chemically defined media for production of insect cells and viruses *in vitro, Dev. Biol. Stand.*, 46, 29, 1980.
34. Sarver, N. and Stollar, V., Sindbis virus-induced cytopathic effect of *Aedes albopictus* (Singh) cells, *Virology*, 80, 390, 1977.
35. Peleg, J. and Pecht, M., Adaptation of an *Aedes aegypti* mosquito cell line to growth at 15°C and its response to infection by Sindbis virus, *J. Gen. Virol.*, 28, 231, 1978.

36. Bhat, U. K. M. and Singh, K. R. P., Structure and development of vesicles in larval tissue culture of *Aedes aegypti* (L.), *J. Med. Entomol.,* 6, 71, 1969.

37. Bhat, U. K. M. and Singh, K. R. P., Establishment of a diploid cell line from larval tissues of *Aedes vittatus* (Bigot, 1861), *Curr. Sci.,* 39, 388, 1970.

38. Sweet, B. H. and McHale, J. S., Characterization of cell lines derived from *Culiseta inornata* and *Aedes vexans* mosquitoes, *Exp. Cell Res.,* 61, 51, 1970.

39. Igarashi, A., Isolation of a Singh's *Aedes albopictus* cell clone sensitive to dengue and chikunguna viruses, *J. Gen. Virol.,* 40, 531, 1978.

40. Deubel, V. and Digoutte, J. P., Morphogenesis of a yellow fever virus in *Aedes aegypti* cultured cells. I. Isolation of different cellular clones and the study of their susceptibility to infection with the virus, *Am. J. Trop. Med. Hyg.,* 30, 1060, 1981.

41. McIntosh, A. H. and Rechtoris, C., Colony formation and cloning in agar medium, *In Vitro,* 10, 1, 1974.

42. McIntosh, A. H., Agar suspension culture and cloning of invertebrate cells, in *Invertebrate Tissue Culture, Research Applications,* Maramorosch, K., Ed., Academic Press, New York, 1976, 1.

43. Peleg, J., Inapparent persistent virus infection in continuously grown *Aedes aegypti* mosquito cells, *J. Gen. Virol.,* 5, 463, 1969.

44. Cahoon, B. E., Hardy, J. L., and Reeves, W. C., Initiation and characterization of a diploid cell lines from larval tissues of *Aedes dorsalis* (Meigen), *In Vitro,* 14, 255, 1978.

45. Varma, M. G. R., Pudney, M., and Leake, C. J., Cell lines from larvae of *Aedes (Stegomyia) malayensis* Colless and *Aedes (s) pseudoscutellaris* (Theobald) and their infection with some arboviruses, *Trans. R. Soc. Trop. Med. Hyg.,* 68, 374, 1974.

46. Bhat, U. K. M. and Guru, P. Y., *Aedes novalbopictus:* establishment and characterization of larval cell lines, *Exp. Parsitol.,* 33, 105, 1973.

47. Singh, K. R. P. and Bhat, U. K. M., Establishment of two mosquito cell lines from larval tissues of *Aedes w-albus, Experientia,* 27, 142, 1971.

48. Marhoul, Z. and Pudney, M., A mosquito cell line (Mos. 55) from *Anopheles gambiae* larvae, *Trans. R. Soc. Trop. Med. Hyg.,* 66, 183, 1972.

49. Schneider, I., Establishment of three diploid cell lines of *Anopheles stephensi* (Diptera, Culicidae), *J. Cell Biol.,* 42, 603, 1969.

50. Pudney, M. and Varma, M. G. R., *Anopheles stephensi* var. *mysorensis:* establishment of a larval cell line (Mos. 43), *Exp. Parasitol.,* 29, 7, 1971.

51. Chao, J. and Ball, G. H., The establishment of a *Culex pipiens* cell line and its adaptation to grow in three different insect tissue culture media, *In Vitro,* 8, 406, 1973.

52. Schneider, I., Establishment of cell lines from *Culex tritaeniorhynchus* and *Culex salinarius* (Diptera: Culicidae) in *Proceedings Third International Colloquium on Invertebrate Tissue Culture,* Slovak Academy of Sciences, Bratislava, Czechoslovakia, 1973, 121.

53. Chao, J. and Ball, G. H., A comparison of amino acid utilization by cell lines of *Culex tarsalis* and of *Culex pipiens,* in *Invertebrate Tissue Culture, Applications in Medicine, Biology and Agriculture,* Kurstak, E. and Maramorosch, K., Eds., Academic Press, New York, 1976, chap. 23.

54. Oro, G., Establishment of a mosquito cell line from *Haemagogus equinus* larvae, *In Vitro,* 20, 153, 1984.

55. Kuno, G., A continuous cell line of a nonhematophagous mosquito, *Toxorhynchites amboinensis, In Vitro,* 16, 915, 1980.

56. Kuno, G., A method for isolating continuous cell lines from *Toxorhynchites amboinensis* (Diptera: Culicidae), *J. Med. Entomol.,* 18, 140, 1981.

Chapter 3

PREPARATION AND MAINTENANCE OF ARTHROPOD CELL CULTURES: ACARI, WITH EMPHASIS ON TICKS

C. E. Yunker

TABLE OF CONTENTS

I. INTRODUCTION

Just as development and virological applications of insect-tissue culture have lagged behind those of mammalian tissue culture, progress in tick-tissue culture has proceeded at a slower pace than in the former endeavor. The sequential development of continuous cell lines from vertebrate, insect, and tick were widely separated events. Likewise, the first indication that animal viruses might be cultivable in explanted mammalian tissue,[1] Trager's proof of arboviral replication in surviving mosquito tissues,[2] and Rehacek's demonstration of tick-borne virus multiplication in primary tick-cell cultures[3] were each separated by spans of 25 years.

In 1952, Weyer attempted to effect transfer of typhus ricksettsie from excised guts of infected lice to uninfected tissues of *Rhipicephalus bursa* ticks by cocultivation in a hanging-drop chamber.[4] The culture medium consisted of rabbit plasma and extracts of lice, rabbit spleen, and rabbit testicle. Although rickettsial transfer was not achieved and no mention was made of the fate of the tick tissues, this attempt represented the earliest reported tick-tissue culture experiment. Weyer concluded that, "Insect tissue, in particular, is ill-suited to in vitro culture, because insects are highly specialized animals, with limited regeneration capacity." Indeed, the concept of embryonic determination of arthropod cells, which implied little or no mitotic activity of explanted tissues, probably served to discourage many who would otherwise have contributed to the field of arthropod tissue culture.

In 1958, Rehacek transplanted tissues of developing adult *Dermacentor marginatus* ticks to hanging-drop slides in various combinations of media and observed fibroblastic outgrowth for 6 days.[5] This observation led him to conduct further experiments designed to improve the condition for in vitro growth of tick tissues and cells and marked the true beginning of tick-tissue culture. In the ensuing years refinements of techniques resolved many technical difficulties associated with the production of adequate amounts of tick cells and, in 1975, resulted in the establishment of the first continuous line of tick cells.[6] Subsequently a number of serially propagating cell lines from various genera and species of ticks have been reported (Table 1). A number of reviews adequately discuss the historical development of tick-tissue culture.[7-9,35,54,60] This chapter will present current methodologies for the production and maintenance of primary tissue and cell cultures and continuous cell lines from ticks. Applications of these in vitro systems to the study of arboviruses are reviewed in Volume I, Chapters 7, 9, and 10 and Volume II, Chapter 13.

II. SOURCE OF MATERIALS

The need for a continuing supply of acarine tissues is best met by the maintenance of laboratory colonies rather than by the collection of materials from the field. In addition, such colonies may sometimes be manipulated to produce aseptic specimens. Because of diverse and complicated life cycles, often involving prolonged feeding on a series of vertebrate hosts, methods for rearing and mass colonization of Acari are not standardized. A discussion of individual methods is beyond the scope of this chapter. Instead, the reader is referred to publications describing rearing and colonization procedures that have proven useful for certain species and groups of these arthropods.

A. Ixodidae

Ixodid or hard-shelled ticks, undergo a single larval and nymphal stage before molting into the adult. The immature stages and females require a protracted blood-meal after which they may leave the host to molt (immatures) or oviposit (females). The majority of the Ixodidae are three-host species that require a different vertebrate for

Table 1
CONTINUOUS TICK CELL LINES CURRENTLY UNDER CULTIVATION[a]

Designation	Species	Source of explant	Ref.
RA 243	*Rhipicephalus appendiculatus*	Metamorphosing nymph	6
ATC-307,-308	*R. sanguineus*	Metamorphosing nymph	30
ATC-304,-310	*Haemaphysalis obesa*	Metamorphosing nymph	30
ATC-281/282,-309	*H. spiningera*	Metamorphosing nymph	30
IX	*Boophilus microplus*	Embryonated eggs	34
VIII-633,-SCC	*B. microplus*	Embryonated eggs	65
BME 26	*B. microplus*	Embryonated eggs	70
RML-15,-18,-19,-20	*Dermacentor variabilis*	Embryonated eggs	59
RML-16,-17	*D. parumapertus*	Embryonated eggs	59
ANE 58	*D. (Anocentor) nitens*	Embryonated eggs	66
RAE 25	*R. appendiculatus*	Embryonated eggs	48
RSE 8	*R. sanguineus*	Embryonated eggs	48
RML-21,-22	*R. sanguineus*	Embryonated eggs	60

[a] This list does not include cell lines lost subsequent to initial report of establishment.

each feeding (e.g., most *Dermacentor* and *Amblyomma* spp.). The two-host species (e.g., some *Rhipicephalus* and *Hyalomma* spp.) remain on the host as immatures but seek out a second host as adults. The one-host ticks (e.g., *Boophilus* spp. and some *Dermacentor* spp.) remain on the same host animal throughout their active stages.

For many years at the Rocky Mountain Laboratory, Hamilton, Mont., a three-host species, *D. andersoni,* was colonized in large numbers in connection with the study of Rocky Mountain spotted fever and the manufacture of a tick-based vaccine against the organism. The methods of colonization were described in detail by Kohls in 1937 and remain essentially unchanged today.[10] They are generally applicable to a number of three-host species and, with modification, to those with differing requirements. Gregson[11] and Philip[12] review similar techniques.

Briefly, adult ticks are fed in capsules taped to the clipped abdomen of a rabbit, while larvae and nymphs are shaken onto the ears and head of a different rabbit, which is housed in a tightly bagged cage. The ticks will detach from the host after engorgement. In the presence of males *D. andersoni* females will feed to repletion in 8 to 10 days, and larvae and nymphs require 5 and 8 days, respectively. Replete ticks are sorted free of fecal material, washed, and stored individually in shell vials (female) or in large quantities in glass cylinders capped at either end with muslin. Molting and oviposition occur in these chambers, which are held in desiccator jars at specified temperatures (e.g., 22 to 30°C) and adequate humidity (e.g., 93% relative humidity [RH]). Temperature markedly affects the development of nonfeeding stages. Unfed, recently molted forms may be stored for many months at low temperatures (e.g., 15°C), provided the RH is maintained at 80%. However, fed immatures succumb to molds at this RH while tending to desiccate at lower values. Metamorphosing nymphs, which are excellent donors of tissue culture materials, may be maintained in an optimal condition for explantation over many weeks if stored in the freezing compartment of a household refrigerator.[13] Engorged females of at least one species, *Amblyomma americanum,* may be kept in cold storage for a number of weeks in order to defer oviposition.[14] Fed females should be surface-sterilized (see below) if their eggs are to serve as tissue culture materials. Individual methods for the rearing or mass production of selected ixodid species have been published: *Ixodes dammini,*[15] *Amblyomma* spp.,[16-18] *Hyalomma* spp.,[18,19] and *Rhipicephalus* spp.[18,21]

B. Argasidae

Like ixodids, most argasids, or soft-shelled ticks, require vertebrate blood-meals in each stage. However they differ biologically in a number of respects. Whereas larvae of some species may complete their feeding in minutes or not feed at all, this stage typically remains attached to a host for long periods of time; nymphs and adults often complete their feeding in a matter of minutes or hours and leave the host. Nymphs may feed several times while undergoing a succession of preadult molts. Adults typically take multiple blood-meals, with females leaving the host to deposit small numbers of eggs after each feeding. However, some species (e.g., *Ornithodoros lahorensis* and *Otobius megnini*) may remain on the same host animal for all or part of their active lives. As with ixodids, temperature is an important regulator of development. Acceleration of oviposition and embryogenesis, as well as a reduction in number of nymphal molts, may be obtained with higher temperatures.[22] Micks[23] and Kaiser[24] describe rearing and colonization techniques for bird parasitizing *Argas* spp. and Gregson[11] reviews methods applicable to mammal parasites.

C. Mites

The life cycles of some mites are equally as complex as those of ticks. Parasitic forms (e.g., Laelapidae, Macronyssidae, and Trombiculidae) have been successfully colonized.[12,25,26] Free-living mites implicated in house-dust atopies (e.g., Acaridae or Pyroglyphidae) are readily produced in mass culture, but production of aseptic colonies requires special procedures.[27-29]

III. PRIMARY CULTURES FROM ORGANS, TISSUES, AND CELLS OF TICKS AND MITES

A. Surface Sterilization of Ticks

The exterior of ticks is readily decontaminated with chemosterilants if precautions are taken to remove dried blood, feces, host hair, and scales. This may be done by washing the ticks in running tap water for several minutes, or by briefly shaking them in 70% ethanol, acetone, surfactant, or detergent solutions, or in 3% hydrogen peroxide. Rubbing the tick between layers of a sterile gauze pad soaked in one of these fluids is also helpful. Following this a number of procedures for surface decontamination may be used. These include washing or soaking ticks for several hours in 0.0002 *M* merthiolate solution,[12] for 15 min in White's solution,[30,31] or, most commonly, in a 0.5 to 1.0% solution of benzalkonium chloride.[6,32,33] A wash or short soak in 70% alcohol usually follows the treatment with benzalkonium chloride. Thereafter, the ticks are rinsed free of the chemicals in sterile distilled water to which antibiotics and antimycotics may be added.[30,31,34] Fungal contaminants may sometimes be eliminated by immersing the ticks in a 0.5% solution of household bleach before treatment with benzalkonium chloride.[35]

B. Tick-Tissue Culture Medium

Early attempts to grow tick organs, tissues, and cells in vitro employed a wide variety of media. Most of these were empirically devised[5,8,36-39] or tried because of their beneficial effects on other cells, vertebrate or invertebrate. One medium specificially formulated for tick cells was based on an analysis of tick hemolymph (*Boophilus microplus*,[40] but this has been only rarely employed.[57] Likewise those devised for insects (Grace's medium,[41,42] Vago and Chastang's medium,[43,44] and Mitsuhashi-Maramorosch's medium[45]) either fail to support growth of tick cells or are only marginally suitable. Currently the most satisfactory media for tick-tissue culture are those formulated for vertebrate cells and commercially available (i.e., lactalbumin hydrolysate medium in Hank's base, HLH,[36,47] Leibovitz's L-15 medium,[6,32] and L-15 combined with Eagle's minimal essential medium [EMEM] in Hank's salts[33]). These media, although far from optimal[35] (see also Chapter 7 of this volume), do permit growth and establishment of tick cell lines. They must be used with essential supplements, bovine plasma albumin, and tryptose phosphate broth,[48] as well as with heat-inactivated fetal bovine serum (FBS). The pH of the medium is critical (the optimal range is 6.6 to 6.8 in most cases) and commercially available dry powders, which omit bicarbonate, are preferable to the same media sold in liquid form. The growth-promoting effects of tick egg extract (TEE), delipidized protein of TEE, and proline on various continuous lines of tick cells have been reported.[48,49]

1. Preparation of Tick-Tissue Culture Medium

The following formulation was used during the establishment of *Dermacentor* and *Rhipicephalus* embryonic cell lines[31,33,59,60] and is also adequate for primary cultures of tick viscera. The formula given here will make nearly 6 L of complete medium. Mix 18 g tryptose phosphate broth (TPB) powder (No. 0080-02-5, Difco Labs., Detroit, Mich.) in 600 mℓ triple glass-distilled or Type I reagent grade ("polished") water; autoclave and allow to cool. Prepare antibiotic stock solutions as follows: penicillin G, 50,000 U/mℓ; streptomycin sulfate, 50,000 μg/mℓ; neomycin sulfate, 10,000 μg/mℓ. Thaw enough FBS to provide 1140 mℓ; inactive at 56°C in water bath for 1 hr. Stir two liter-sized packets of Liebovitz's L-15 medium with glutamine (No. 430-1300, GIBCO, Grand Island, N.Y.) in 2 ℓ sterile, triple glass-distilled or polished water. Similarly, mix 2 packets of Eagle's minimal essential medium with Hank's salts and glutamine in 2 ℓ sterile, triple glass-distilled or polished water. Do not add the bicarbonate called for on packet. Combine with L-15 medium and stir briefly. Add 1140 mℓ FBS, 570 mℓ TPB, and 5.7 g bovine plasma albumin (BPA) fraction V (No. 2293-01, Armour Biochemical, Los Angeles, Calif.). Add 11.5 mℓ each of penicillin G and streptomycin sulfate stock solutions, and 23 mℓ neomycin sulfate stock solution. (Final concentrations of additives in complete medium are: FBS, 20%; TPB sol., 10%; BPA, 0.1%; penicillin G, 100 U/mℓ; streptomycin sulfate, 100 μg/mℓ; neomycin sulfate, 40 μg/mℓ.) The final pH of medium should be 6.6 to 6.8; adjust if necessary with 1 N HCl or 1 N NaOH. Sterilize by pressure filtration through a cellulose acetate filter, discarding the first 100 mℓ; dispense remainder in sterile, 250 mℓ Wheaton bottles; fill to neck of bottle and cap tightly; refrigerate at 4°C. Medium should not make contact with bottle cap liners.

Because of a unique sensitivity of tick cells, particularly in early passages, special precautions must be used to exclude toxic substances. These substances originate in a variety of sources, notably water and biological components of the medium. It is necessary to use water of high quality for preparation of solutions and to pretest all additives before routine use. Polished water (i.e., that produced by mixed-bed deionization and carbon adsorption of previously distilled or deionized water) yields consistently good tissue culture media, but is not without disadvantages. This water, of unusually high purity, is expensive to produce, cannot be stored in glass containers for any length of time due to its leaching effect, and may inhibit in some manner the regularity of viral plaque formation. Toxic FBS is the most common cause of tick tissue loss and cell cultures. Each lot of FBS must be pretested by adding 10% new FBS to the maintenance medium (i.e., above medium prepared with only 10% FBS of proven quality). This trial medium is introduced into one or two flasks of tick cells, which are observed daily. Toxic effects are visible as early as 2 hr and as late as 14 days after introduction of the medium. They are seen as contraction and rounding of cell membranes and detachment of cells from the substrate. Also, the filter pad should be washed by passing through it large quantities (e.g., 2 to 5 ℓ) of warm triple glass-distilled or polished water before use. All glassware, including medium storage bottles, should be washed and rinsed thoroughly and kept separate from common glassware.

C. Primary Cultivation of Tick Organs, Tissues, and Cells

Primary cultures intended for virus inoculation are often prepared from engorged nymphal ticks held until the developing imaginal tissues are visible within the nymphal integument. These tissues, as well as individual organs (including midgut, salivary glands, ovaries, Malpighian tubules, and hypodermis), may be transplanted to culture vessels intact or as partially dissociated tissues[5,30,36,38,44,47,54] or as dispersed cells.[3,6,50-52] Adult tick organs and tissues are less often employed, assumably because most cells of this stage have largely lost the potential to divide. Two exceptions to this are ovaries[8,54] and hemocytes[55,56] which produce adequate numbers of cells. Ovarian cell cultures have been successfully subdivided, but hemocytes usually die when treated so. Primary cultures of embryonic cells would conceivably be useful in arbovirus studies, but have found widest application in the initiation of continuous cell lines.[32-34,48,59]

1. Preparation of Primary Cultures from Nymphal Ticks (Dermacentor andersoni)

Nymphal ticks[47] are held at 26 to 28°C for at least a week after dropping off the host. By this time the adult stage has begun to take form and tissues are amenable to transplantation. Engorged ticks are satisfactory explant donors for as long as 2 weeks, after which hardening of the adult cuticle signals a lessened ability of the explant to yield cellular outgrowth. Externally sterilized ticks are fastened to wax-filled dissecting dishes with small entomological pins (*minuten nädeln*), or to glass microscope slides by means of glue or double-coated cellophane tape. All surfaces and instruments are sterilized; exposure to ultraviolet light is adequate for microslides, glue-nozzles, and tape, while immersion in 70% alcohol or flaming is used for the instruments and wax-filled dishes. Three shallow depressions are formed in the surface of the wax into which a tick, dorsal side uppermost, is fastened with two *minuten nädeln*. The tick is opened with iridectomy scissors or a microscalpel by a circumferential incision around the periphery of the body. The dorsal integument is teased away from muscles and tracheal branches with watchmaker forceps and is placed in a second well containing tick-tissue culture medium or Hank's balanced salt solution (HBSS). The "backless" tick is gently washed several times in medium or HBSS by means of a Pasteur pipette, and imaginal tissues (or individual organs) are removed to a third well containing medium or HBSS. Washing solutions should contain a 10 × concentration of antibiotics. The alimentary tract and Malpighian tubules may be separated from the imaginal tissues or retained. Exclusion of these nymphal components greatly improves appearance of the cultures without significant reduction of initial cellular outgrowth, but does reduce longevity of the cultures.[38] Their removal is recommended if explants are to be dissociated and prepared as monolayers of dispersed cells. After removal of the viscera the hypodermis may be scraped from the dorsal and ventral walls of the integument and pooled with the material to be explanted; 5 to 10 nymphal ticks provide enough material to seed a single, 25 cm² plastic tissue culture flask. Fewer nymphs or individual organs may be placed in plastic culture tubes. The 16 × 110 mm Lux Ambitube (No. 5102, Miles Laboratories, Naperville, Ill.) is especially useful in that it has a flattened side of good optical clarity. A grid-like pattern may be scratched into the floor of plastic culture vessels by means of a bent steel probe in order to enhance attachment of tissues. Whole viscera or organs are lightly centrifuged in two washes of medium and added to vessels in a small amount of medium with a Pasteur pipette. Vessels are allowed to stand undisturbed for 1 hr, after which culture medium containing unit concentrations of antibiotics are added (0.70 m*l*/tube,4 m*l*/25 cm² flask). Cultures are incubated at 26 to 28°C in a humidified atmosphere, or at higher temperatures for shorter periods if arboviruses are introduced. Both uninfected and infected explants will produce outgrowths of fibroblastic cells within hours of explantation. Outgrowths will increase in size for as long as 1 to 2 months, after which the frequency of appearance of new fibroblast-like cells decreases. The explants may then remain static for a number of weeks, outgrowths may sporadically reappear or, commonly, cellular degeneration may ensue with the loss of the culture. Individual cultures of whole imaginal tissues may remain viable for long periods (e.g., 5 months for *Dermacentor andersoni* and 9 months for *Rhipicephalus appendiculatus*).[47]

Isolated organs of ixodid ticks have been transplanted to culture vessels a number of times since the original experiments of Weyer,[4] where they have served as substrates for the growth of arboviruses, rickettsias, and protozoans.[8] Cultured reproductive organs of either sex, intestinal tracts, salivary glands, Malpighian tubules, connective tissue, and hypodermis have survived for various lengths of time, ranging from 1 week to over 2 $1/2$ months.[9,63] Recently, explanted midguts and salivary glands of *D. albipictus* females have been shown to survive in vitro, as determined by contractions or outgrowths of fibroblast-like cells, respectively, for over $6^1/_2$ months.[67]

Metamorphosing nymphal tick viscera or individual organs may also be dissociated, enzymatically or mechanically, and explanted as dispersed cells which yield nearly confluent monolayers. This method devised by Rehacek[50,52] and later modified to produce more uniform cell sheets[53] is useful in assessing the growth of arboviruses in primary monolayers of tick cells and also provides cultures amenable to serial passage.[6,8,54] For example, digestive tract, rectal sac, and Malpighian tubules are dissected from 8 to 10 ticks and discarded. The remaining tissues are collected and pooled as described above, washed a number of times in diluent containing extra antibiotics, and minced with small scissors in 0.25% trypsin solution in 1:5000 versene.[58] The minced tissues are suspended in 8 to 10 mℓ trypsin solution and held for 10 min at room temperature. Cells are then dissociated by vigorous pipetting with a Pasteur pipette and collected by low speed centrifugation. The pellet is dispersed in saline and cells are centrifuged and resuspended in 1 mℓ of medium. This quantity provides an adequate seeding density for one Leighton tube, whereas 4 mℓ of suspension are required to seed a 25 cm^2 plastic flask.

A modification of this technique omits trypsinization.[30,54] This was deemed necessary because the enzyme was not only detrimental to tick cells, but resulted in the loss of much tissue. Pooled tissues are teased apart with fine forceps and the fragments are seeded directly into the container. This method provides a larger quantity of tissues for primary cultures and is also said to shorten the interval between seeding and first subculture.

Early attempts to cultivate tick hemocytes in vitro resulted in only brief survival of the cells,[3,8] however, these experiments demonstrated that tick cells were actively phagocytic and capable of supporting arbovirus growth. Later a technique was devised for collecting and culturing larger numbers of hemocytes from ixodid ticks (*D. andersoni*).[55] These cells survived for at least 71 days, as judged by their ability to support growth of Colorado tick fever virus. Here, female ticks are fed to near repletion, held for 3 to 4 days at ambient conditions, and then stored at 4°C over a saturated solution of ammonium sulfate. Under these conditions the ticks are satisfactory donors of hemocytes for up to 3 months. The external surfaces of the ticks are disinfected and fastened to microscope slides with double-surfaced cellophane tape or glue. Hemolymph is drawn into a 0.4 mm (ID) capillary tube from a short shallow incision made just posterior to the scutum (Figure 1). After the incision is made, very slight downward pressure is applied to the posterior of the tick until hemolymph begins to exude. The capillary tube is applied to the welling hemolymph and slight pressure is continued. Excessive pressure invariably causes the gut or Malpighian tubules to extrude, blocking

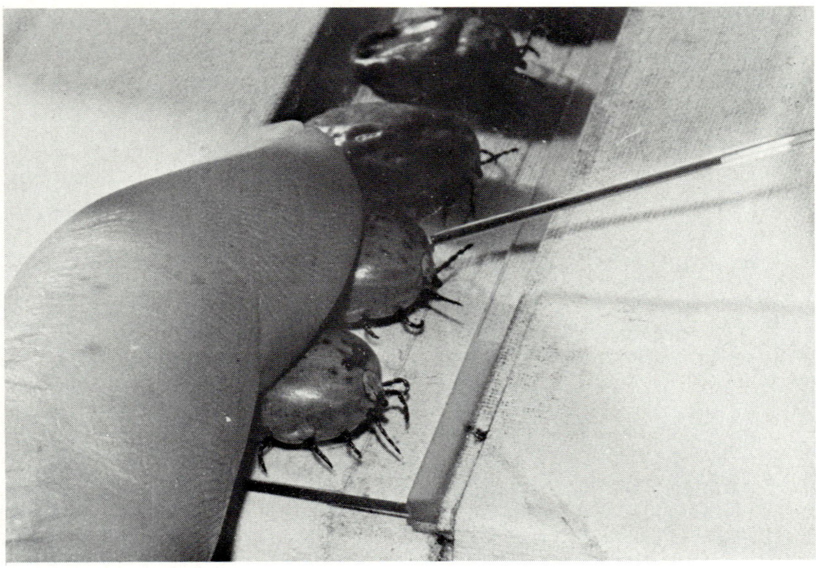

FIGURE 1.   Capillary tube collection of hemolymph exuding from an incision made posterior to scutum of a partially engorged female of *Dermacentor andersoni*. (Courtesy of Entomological Society of America.)

the incision. Precautions are taken to prevent contamination of the capillary tube, which is filled to $^1/_3$ of its length and placed in a 16 × 100 mm, screw-capped tissue culture tube containing sufficient culture medium to cover the capillary tube. In each culture tube four of the latter are placed in it, and it is then tightly capped and incubated at 27°C in a horizontal position. The hemolymph containing cells may also be expressed with a rubber bulb into culture medium in micro-culture vessels, such as 96-well tissue culture plates or Rose chambers (Figure 2). Care must be taken to transfer cells immediately from capillary tube to tissue culture well because of the marked tendency of hemocytes to rapidly adhere to tube surfaces. Cultures are incubated for 4 to 14 days before inoculation, and contaminated ones, or those with low cell populations, are discarded. Tube cultures are centrifuged at 140 × g for 10 min; existing medium is replaced with that containing virus-inoculum and cultures are reincubated. Temperatures higher than ambient (e.g., 34 to 37°C) are usually preferred for adsorption of arboviruses and, if not prolonged, are not detrimental to the tick cells. Cultures may be sampled at intervals and medium changed weekly after centrifugation as above. No more than $^3/_4$ of the existing medium is removed at each weekly change. Cells of mixed morphology (fibroblastic, spindle-shaped, rounded) will attach as a monolayer and retain a healthy appearance for a number of weeks (Figure 3). A gradual deterioration in appearance and decline in the number of cells will ensue. Attempts at subcultivation of these cells generally results in loss of the culture. We have cultured hemocytes from argasid ticks (*Argas [Persicargas] arboreus*) in Rose chambers, capillary tubes, and small (2 dram) screw-cap glass vials, obtaining confluent monolayers of healthy appearing cells (Figure 4) which persisted for over 2 months.[68] As with ixodid hemocytes, the cells failed to survive subculture attempts, however, hemocytes from other argasid ticks (*Ornithodoros coriaceus*) cultured by the above method for 6 weeks were said to grow out of the capillary tube opening onto the surface of the culture flask.[56] This fact, and the establishment of continuous lines of embryonic tick cells resembling hemocytes (Figures 5 and 6),[60] suggests that serially propagating lines from tick hemocytes are possible.

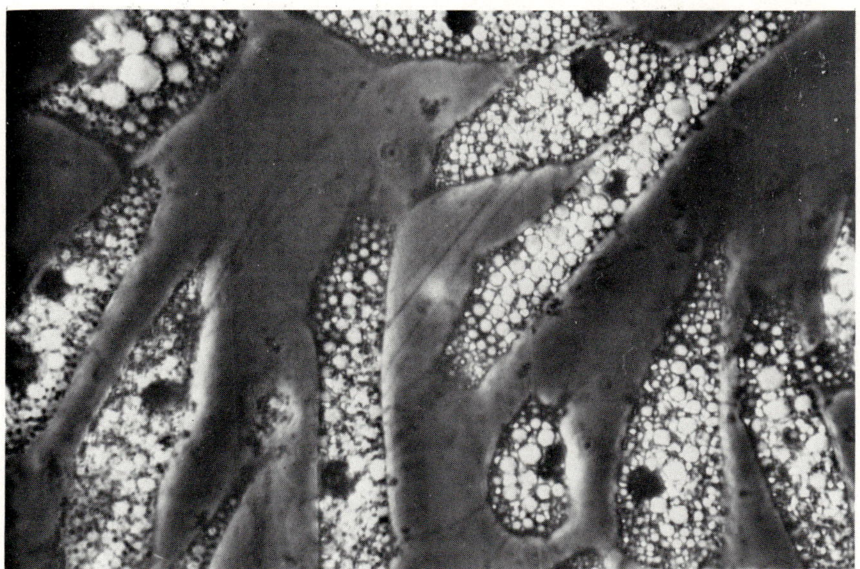

FIGURE 2.   *Dermacentor andersoni* hemocytes in a Rose chamber culture, 72 hr after explantation.

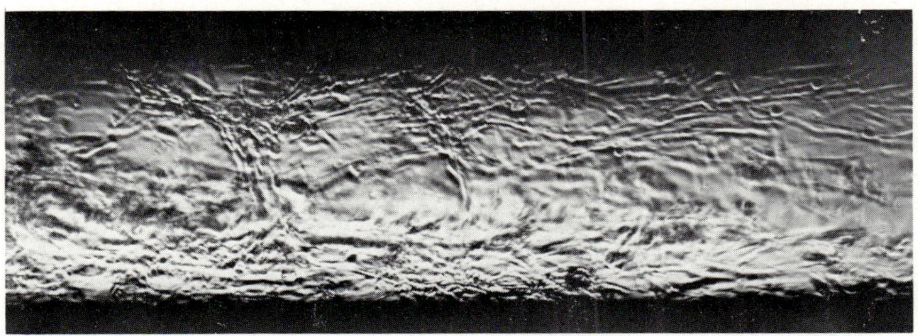

FIGURE 3.   *Dermacentor andersoni* hemocytes in capillary tube culture, 37 days after explantation (magnification × 875).

2. Primary Cultivation of Mite Tissues

Apparently only one attempt at cultivation of mite tissues has been reported.[27,71] The species involved was *Dermatophagoides farinae* (Family Pyroglyphidae), the American house-dust mite, a cosmopolitan, free-living house inhabitant associated with house-dust allergies. Mites and eggs were collected from axenic mite-cultures and surface-sterilized by slowly pouring sterilizing solution over them for 3 min as they were held in a paper funnel. Benzethonium chloride (Hyamine®, Lonza Corp., Fairlawn, N.J.), as a 0.1% solution, was selected as a routine sterilant because it permitted the greatest survival of mites. The sterilant was removed by a 2 min wash in 50 m*l* of distilled water containing a drop of surfactant, Triton X-100® (Emulsion Engineering, Inc., Elk Grove Village, Ill.). Cells obtained from these mites were placed in hanging-drop cultures in Yunker, Vaughn, and Cory's medium,[61] and were seen to spread out over the surface of the cover glass. The optimal osmotic pressure of the medium was determined to be 0.684 Osm. A majority of cells survived for 5 days, with some cells persisting as long as 9 days. A photomicrograph of attached cells shows healthy-ap-

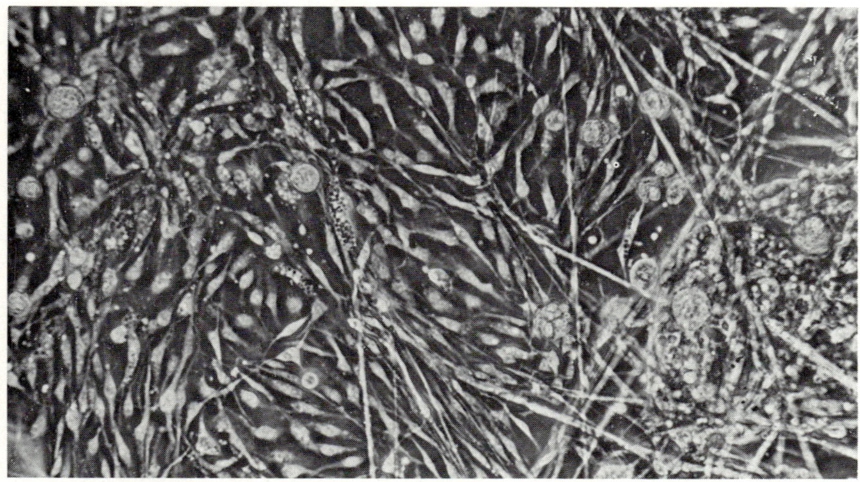

FIGURE 4.   *Argas (Persicargas) arboreus* hemocytes cultured in glass vial, 41 days after explantation.

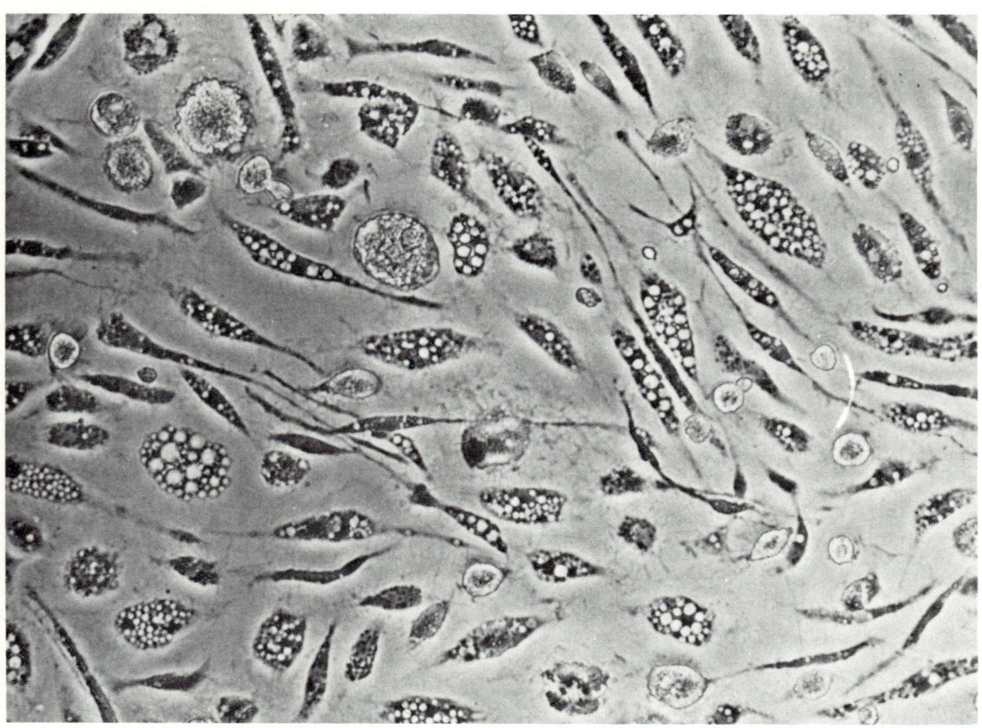

FIGURE 5.   RML-22 cell line established from embryos of *Rhipicephalus sanguineus,* 36th passage, approximately 400X. Compare with Figure 2. (Courtesy of Ellis Harwood, Ltd., Chichester, England.)

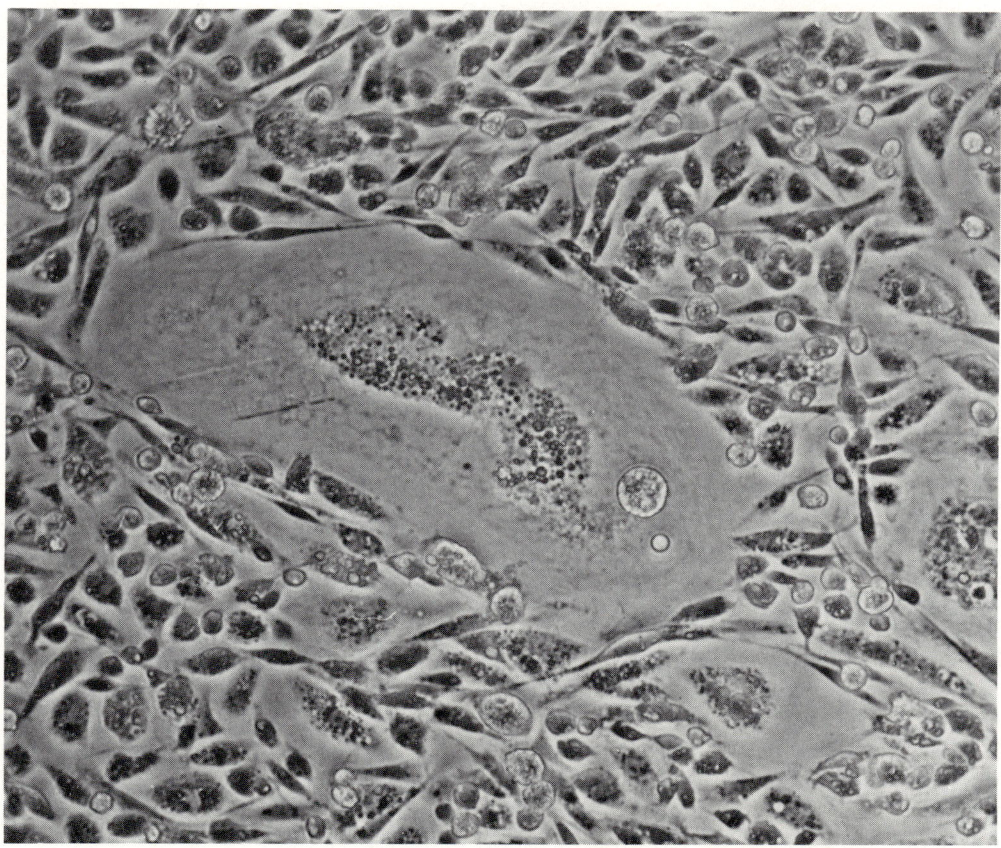

FIGURE 6.   RML-21 cell line established from embryos of *Rhipicephalus sanguineus,* 28th passage (magnification approximately × 400).

pearing, multipolar fibroblastic cells. Hence, additional efforts to culture mite cells would appear promising.

IV. CONTINUOUS CELL LINES FROM TICKS

A. Initiation and Early Maintenance of Cultures

All established tick cell lines, regardless of tissue of origin, have evolved from primary cultures prepared from dissociated tissues or dispersed cells, rather than from explants of whole organs or organ systems. The techniques for preparing such primary cultures from metamorphosing nymphs is described above; procedures for preparation of tick embryo cell cultures[58,59] are reviewed here. Maintenance and subculture of both systems are similar, and procedures follow below.

Female ticks fed in the presence of males are collected soon after dropping off the host and surface sterilized (see Section III A). After a final rinse with sterile distilled water, they are allowed to dry and are placed individually in sterile, cotton-stoppered glass vials, each containing a strip of sterile filter paper slightly moistened with 10 × antibiotic-antimycotic solution (No. 600-5245, GIBCO Labs., Grand Island, N.Y.). Females are held at 80% RH and 27°C until oviposition has peaked, after which they are removed from the vials and discarded. Egg masses from 2 to 4 ticks are aseptically collected 8 to 12 days after average onset of oviposition and pooled. Eggs are dispersed in acetone, briefly washed twice in this fluid, and then soaked for 10 min in White's solution (0.25 g $HgCl_2$, 6.5 g NaCl, 1.25 mℓ HCl, 250 mℓ 95% ethyl alcohol, and 750 mℓ H_2O). They are then freed of this solution by numerous rinses in sterile distilled water and transferred to a sterile, 100 mm petri dish. Final rinse-water is removed with a Pasteur pipette and 2 to 3 mℓ culture medium is added. Eggshells are cracked by applying light pressure with a flat end of a glass or lucite rod. Care must be taken to avoid crushing the embryos. Enough medium is added to bring the volume to 10 mℓ and, with a 12 mℓ syringe and 16 gauge, $1^1/_2$-in. needle, embryonic tissues are gently worked into a dispersed suspension by repeatedly evacuating and refilling the syringe. The suspension is then drawn into the barrel of the syringe and the needle is replaced with a sterilized filter-holder containing a disc of 16XX bolting silk. Syringe contents are gently expressed into a 15 mℓ centrifuge tube and washed twice in culture medium by centrifugation at $140 \times g/10$ min. The final supernatant is decanted and cells are resuspended in 8 mℓ medium. This suspension is divided equally between two 25 cm^2 plastic tissue culture flasks that have been conditioned by incubation with 5 mℓ growth medium at 37°C/24 hr. (This medium is discarded before flasks are seeded.) Flasks are tightly capped and incubated at 27°C.

Cells begin to attach to the flask surface within hours after seeding, but attachment may not be completed for several days. In the weeks following attachment, cellular outgrowths can be seen surrounding undissociated clumps of cells and tissues. These clumps consist mostly of fibroblastic cells that sometimes produce hollow, floating, multicellular vesicles. Thereafter, the cellular outgrowths increase in size and number of cells. Often, thickened foci of cells arise from these outgrowths, but confluent monolayers are seldom seen in early primary cultures. A replacement of large, granular fibroblastic cells with patches of small, clear round cells at about 4 weeks may be noted,[58] and the latter cells may proliferate to nearly (or completely) cover the flask surface. At this stage the cells may be subcultured or, if not completely confluent, "reseeded".[6] These procedures are accomplished after replacement of one half of the existing medium with fresh medium by detaching the cells from the flask surface by means of a Tygon® or rubber policeman and dispersing them in the flask by gentle pipetting with a Pasteur pipette. The cells are allowed to reattach (reseeding), or one half volume of suspension may be transferred to a conditioned flask and the volume of medium in both flasks brought to 4 mℓ with fresh medium. Reseeding liberates cells from congested clumps and allows them to reattach more uniformly in the original flask. In early subcultivations use of trypsin is avoided, but in later passages tick cells are not adversely affected by this procedure. Typically, tick cells adapt to the in vitro situation only after a prolonged period in primary culture, and cell growth is extremely slow and erratic after early passages.[9] Only after the rate of passage becomes regular can these lines be considered established. In contrast to mosquito cell lines, this may take from 1 to over 3 years to become evident for tick cell lines.

B. Routine Maintenance of Tick Cell Lines

Cultures are evaluated weekly and either reincubated, given a change of medium, or subcultured. A useful classification system is to grade cultures on a scale of 1 to 4, with plus (+) or minus (−) symbols used to indicate intermediate grades, improvement, or deterioration. Here, 1 = few attached cells; 2 = adequate cells but no confluency; 3 = confluency without crowding; and 4 = a tightly confluent monolayer often with cell clumps or floating cells. Cultures graded 1 − to 2− are maintained by leaving them undisturbed or by removing all but 1 mℓ of existing medium and adding 2 mℓ of fresh medium. For those graded 2 to 3−, the medium is similarly changed but the amount added is increased by 1 mℓ (total volume/25 cm^2 flask = 4 mℓ). Those graded 3 to 4+ are decanted of all medium and floating cells and are subcultured. Occasionally, particularly after transfer, some cells will fail to attach (or spontaneously detach) and deteriorate. Healthy, attached cells should be freed of floating cells and debris by gently washing the monolayer in 3 mℓ of fresh medium. The wash is discarded and fresh medium is restored to the flask.

Cells from established tick lines are generally amenable to dispersal by trypsin and trypsin/versene, and to frozen storage in dimethylsulfoxide (DMSO). Primary cultures and young lines are variable in their response to these treatments, however, the latter cultures and lines may be maintained in viable condition for several months when held at 12°C.[35] Tick cells may also be dispersed for subculture or freezing by policing, pipetting, or by forcefully shaking the flask with short, abrupt motions. These mechanical manipulations invariably result in some cell damage and, in addition, pipetting and shaking tend to select poorly attached cells.

To trypsinize newly formed, confluent monolayers of tick cells, two cultures in 25 cm² flasks are decanted of medium and briefly rinsed with 4 mℓ of 0.25% trypsin (1:250 GIBCO, Grand Island, N.Y.) in Dulbecco's saline (Ca⁺⁺ and Mg⁺⁺ free). An additional 2 mℓ of trypsin solution is added to each flask, which is then incubated at 37°C/5 min, or until the cell sheet begins to slough. Cells are then gently dispersed and pooled in 10 mℓ medium by means of a Pasteur pipette. For subculture, the suspension is divided among the parent flask and 2 to 4 new, conditioned flasks. The volume of medium in each flask is brought to 4 mℓ and the flasks are tightly stoppered and reincubated. For freezing, the cell suspension is centrifuged at 100 × g/10 min, the pellet is resuspended in 1.8 mℓ medium and is transferred to a sterile, plastic freezing vial. DMSO is added dropwise (0.2 mℓ) and the mixture is allowed to equilibrate undisturbed for 20 to 30 min. Vials are slowly frozen (1°C/min) to −20°C. This rate may be approximated by wrapping them in a plastic foam material or in multiple layers of paper towelling and placing them in a −70°C freezer overnight. Later they may be unwrapped and transferred to liquid N₂. To revive cells, ampoules are thawed rapidly in a 37°C water bath. Cells are centrifuged at 100 × g/10 min, supernatant medium is discarded, and cells are resuspended in 5 mℓ fresh medium. This process is repeated and cells are resuspended in 4 mℓ medium, preferably containing 1 mℓ of filtered, spent medium from an ongoing culture. The suspension is transferred to a conditioned 25 cm² flask which is then incubated at 27°C. Within 3 to 24 hr, cells will attach, whereupon medium and unattached cells are decanted and replaced with a 3:1 mixture of fresh:spent medium. The unattached cell suspension may be transferred to a second conditioned flask where attachment will often occur.

C. Characteristics of Continuous Lines of Tick Cells

Kurtti and Munderloh[9] compared morphology, karyotype, and growth rate among a number of cell lines from various genera and species of ixodids. Cultures exhibit a variety of cell types characterized as fibroblastic, epithelioid, or hemocyte-like. A given line, while often heterogeneous in earlier passages, may become predominantly of a single type. Cells may possess either the male or female complement of chromosomes, or both. Although most lines are diploid, aneuploidy is sometimes seen. Growth rates of tick cells, usually incubated at 28°C, are characteristically slow in that they are often split 1:2 or 2:3 on a weekly basis. A line from *Rhipicephalus sanguineus* (RSE 8) grown at 31°C is split 1:5 weekly[9] yielding approximately 40 to 60 times as many cultures over a span of one month. The slow growth of some *Dermacentor variabilis* cells has been accelerated by adaptation to higher temperatures.[59] In fact incubating cultures at slightly higher temperatures may improve chances for continuous cultivation.[9] It has been pointed out that in most cases the procedures used to establish tick cells as serially propagating lines are similar and tend to favor the selection of slow-growing, microaerophilic populations having low plating efficiencies and requiring high cell densities.[9] It was suggested that modification of cultural procedures (e.g., cloning, selecting for attachment, and addition of growth factors) may result in lines possessing entirely different characteristics, especially with regard to their ability to support growth of tick-borne pathogens having high energy requirements.

D. Factors in the Successful Culture of Tick Cell Lines

The reasons for growth or nongrowth of tick cells in culture are not fully understood. However, a body of knowledge has accumulated which, if applied, will increase the likelihood of producing healthy cultures of growing cells.[9,33,58] Some of the factors involved have been mentioned above or are discussed below. Some are not unique to tick cells but pertain to cell culture in general.

1. Avoidance of Contamination

Surface sterilization of ticks and eggs is usually readily accomplished, as outlined in Section III.A. Occasionally contaminating microorganisms are carried internally by ticks, especially in the midgut, which is the organ most likely to succumb to mold or yeast infections. The contaminants are generally seen within 3 to 7 days of explanation. For this reason it is useful to prepare midgut cultures in tubes containing the digestive tract from only one specimen. After incubation for a week, the contamination is usually obvious and affected cultures may be culled and healthy ones pooled, if desired. Valuable cultures such as young cell lines may sometimes be freed of contaminants by washing the cells and adding a 10 × concentration of antibiotic solution to the replenished medium. This is particularly successful in the case of molds which tend to grow in discrete mycelia. The latter may be picked out of the culture with a Pasteur pipette, after which the cells are washed and given fresh medium containing 25 μg/mℓ amphotericin-B (Fungizone®, E. R. Squibb & Sons, Inc., NY).

2. Species of Ticks

Evidence indicates that the cells of some species of ticks are more amenable to cultivation than are those of others.[33] This is most apparent when subcultivation of primary cultures is attempted. The most obvious reason for this is subtle inadequacies of current media, which should be viewed as the next challenge of the tick tissue culturist.

3. Suitability of Tissue

The tissues chosen must not only be capable of growth, but be of the proper physiological age in order to provide an adequate seeding density. The latter requirement is seen in embryonic cell cultures which fail to grow or grow poorly when ixodid egg clusters younger than 1 week are used. The reason for success after this time is probably related to the fact that the nucleic acid content of the developing tick embryo increases most dramatically beginning on postoviposition day 6.[60,64]

Metamorphosing nymphal tissues must also be used at the proper time of development to provide mitotically active cells in adequate number. This is generally agreed to be at such a time as the developing image becomes visible within the nymphal integument.[60] It had been thought that adult tissues developing within the nymphal skin must be used before the adult cuticle begins to form.[35] However, the isolation of cell lines from full-term embryos as well as neonate larvae[48] argues against this.

4. Nontoxicity of the Medium

The exquisite sensitivity of tick cells (particularly of younger cultures) to toxic substances in biological components of the medium has been noted.[33,60] Each new lot of medium additive must be screened before use (see Section IiI.B) and serum must be adequately heat-inactivated before incorporation into the medium. Usually, this inactivation is carried out by incubation at 56°C/1½ hr, but Buckley[69] reports that 37°C/24 hr is adequate. This lower temperature may in fact be preferable in that it would conceivably preserve growth factors in the serum. The need to provide water of the highest quality has been discussed above.

5. Conditions of Cultivation

Both culture medium and culture vessels must be conditioned to allow tick cells to adapt to in vitro conditions. Thereafter, these precautions are often unnecessary. Acid pH in the range 6.6 to 6.8 is a critical condition for tick cells. Undoubtedly, early attempts to cultivate tick tissues and cells were unsuccessful due, in part at least, to the use of alkaline media. A shift of pH from acidity toward neutrality retards multiplication of *Dermacentor* cells without noticeably affecting their appearance, while alkaline pH quickly causes *Boophilus* cells to degenerate.[34,60]

REFERENCES

1. **Steinhardt, E., Israeli, C., and Lambert, R. A.**, Studies on the cultivation of the virus of vaccinia, *J. Infect. Dis.*, 13, 294, 1913.
2. **Trager, W.**, Multiplication of the virus of equine encephalomyelitis in surviving mosquito tissues, *Am. J. Trop. Med.*, 18, 387, 1938.
3. **Rehacek, J.**, Propagation of tick-borne encephalitis (T.E.) virus in tick tissue cultures, *Ann. Epiphyt.*, 14, 199, 1963.
4. **Weyer, F.**, Explantationsversuche bei Läusen in Verbindung mit der Kultur von Rickettsien, *Zentralbl Bakteriol. Parasitenkd. Infektionskr.*, 159, 13, 1952.
5. **Rehacek, J.**, Preliminary report on tick tissue culture, *Acta Virol.*, 2, 253, 1958.
6. **Varma, M. G. R., Pudney, M., and Leake, C. J.**, The establishment of three cell lines from the tick *Rhipicephalus appendiculatus* (Acari: Ixodidae) and their infection with some arboviruses, *J. Med. Entomol.*, 11, 698, 1975.
7. **Yunker, C. E.**, Arthropod tissue culture in the study of arboviruses and rickettsiae: a review, *Curr. Top. Microbiol. Immunol.*, 55, 113, 1971.
8. **Rehacek, J.**, Tick tissue culture and arboviruses, in *Invertebrate Tissue Culture, Applications in Medicine, Biology, and Agriculture,* Kurstak, E. and Maramorosch, K., Eds., Academic Press, New York, 1976, 21.
9. **Kurtti, T. J. and Munderloh, U. G.**, Tick cell culture: characteristics, growth requirements, and applications to parasitology, in *Invertebrate Cell Culture Applications,* Maramorosch, K. and Mitsuhashi, J., Eds., Academic Press, New York, 1982, 195.
10. **Kohls, G. M.**, Tick rearing methods with special reference to the Rocky Mountain wood tick, *Dermacentor andersoni* Stiles, in *Culture Methods for Invertebrate Animals,* Galtsoff, Lutz, Welch, and Needham, Eds., Cornell University Press, Ithaca, N.Y., 1937, 246.
11. **Gregson, J. D.**, Ticks, in *Insect Colonization and Mass Rearing,* Smith, C. N., Ed., Academic Press, New York, 1966, chap. 4.
12. **Philip, C. B.**, Methods of studying ticks and mites as virus hosts and vectors, in *Methods in Virology, Vol. 1,* Maramorosch, K. and Koprowski, H., Eds., Academic Press, New York, 1967, chap. 4.
13. **Yunker, C. E. and Cory, J.**, Effectiveness of refrigerated nymphs in tick tissue culture experiments, *J. Parasitol.*, 51, 686, 1965.
14. **Koch, H. G.**, Cold storage of lone star tick engorged females and eggs, *Southwest. Entomol.*, 6, 140, 1981.
15. **Krinsky, W. L.**, Development of the tick *Ixodes dammini* (Acarina: Ixodidae) in the laboratory, *J. Med. Entomol.*, 16, 354, 1979.
16. **Patrick, C. D. and Hair, J. A.**, Laboratory rearing procedures and equipment for multi-host ticks (Acarina: Ixodidae), *J. Med. Entomol.*, 12, 389, 1975.
17. **Koch, H. G. and Mount, G. A.**, Mass rearing of lone star ticks, *J. Econ. Entomol.*, 73, 580, 1980.

18. Matthewson, M. D. and Hughes, G., Establishment and cultures of two and three-host ticks in the laboratory and their use in the screening of candidate ixodicides, in *Tick-borne Diseases and Their Vectors*, Wilde, J. H. K., Ed., University of Edinburgh, 1978, 231.
19. Weber, G. and Walter, G., A note on the production of unfed nymphs of the tick *Hyalomma anatolicum excavatum* for experimental use, *Ann. Trop. Med. Parasitol.*, 76, 583, 1981.
20. Samish, M., Mass rearing devices for *Hyalomma anatolicum excavatum* (Acari: Ixodidae) on calves and rodents exposed to whole body infestation, *J. Med. Entomol.*, 19, 6, 1982.
21. Bailey, K. P., Notes on the rearing of *Rhipicephalus appendiculatus* and their infection with *Theileria parva* for experimental transmission, *Bull. Epizoot. Dis. Afr.*, 8, 33, 1960.
22. Loomis, E. C., Life histories of ticks under laboratory conditions (Acarina: Ixodidae and Argasidae), *J. Parasitol.*, 47, 91, 1961.
23. Micks, D. W., The laboratory rearing of the common fowl tick, *Argas persicus* (Oken), *J. Parasitol.*, 37, 1, 1951.
24. Kaiser, M. N., The subgenus *Persicargas* (Ixodoidea, Argasidae, *Argas*). 3. The life cycle of *A. (P.) arboreus*, and a standardized rearing method for argasid ticks, *Ann. Entomol. Soc. Am.*, 59, 497, 1966.
25. Audy, J. R. and Lavoipierre, M. M. J., Parasitic mites, in *Insect Colonization and Mass Production*, Smith, C. N., Ed., Academic Press, New York, 1966, chap. 3.
26. Sasa, M., Biology of chiggers, *Annu. Rev. Entomol.*, 6, 221, 1961.
27. Wharton, G. W., House dust mites, *J. Med. Entomol.*, 12, 577, 1976.
28. Rodriguez, J. G., Inhibition of acarid mite development by fatty acids, in *Insect and Mite Nutrition*, Rodriguez, J. G., Ed., North-Holland, Amsterdam, 1972, 637.
29. Rodriguez, J. G. and Blake, D. F., Culturing *Dermatophagoides farinae* in a meridic diet, in *Recent Advances in Acarology*, Vol. 2, Rodriguez, J. G., Ed., Academic Press, New York, 1979, 211.
30. Guru, P. Y., Dhanda, V., and Gupta, N. P., Cell cultures derived from the developing adults of three species of ticks, by a simplified technique, *Indian J. Med. Res.*, 64, 1041, 1976.
31. Yunker, C. E. and Meibos, H., Culture of embryonic tick cells, *TCA Manual*, 5, 1015, 1979.
32. Pudney, M., Varma, M. G. R., and Leake, C. F., Culture of embryonic cells from the tick *Boophilus microplus* (Ixodidae), *J. Med. Entomol.*, 10, 493, 1973.
33. Bhat, U. K. M. and Yunker, C. E., Establishment and characterization of a diploid cell line from the tick, *Dermacentor parumapertus* Neumann (Acarina: Ixodidae), *J. Parasitol.*, 63, 1092, 1977.
34. Holman, P. J. and Ronald, N. C., A new tick cell line derived from *Boophilus microplus*, *Res. Vet. Sci.*, 29, 383, 1980.
35. Kurtti, T. J. and Buscher, G., Trends in tick cell culture, in *Practical Tissue Culture Applications*, Maramorosch, K. and Hirumi, H., Eds. Academic Press, New York, 1979, chap. 23.
36. Martin, H. M. and Vidler, B. O., *In vitro* growth of tick tissues (*Rhipicephalus appendiculatus* Neumann, 1901), *Exp. Parasitol.*, 12, 192, 1962.
37. Kordova, N. and Rehacek, J., Experimental infection of ticks *in vivo* and their organs *in vitro* with filterable particles of *Coxiella burneti*, *Acta Virol.*, 3, 201, 1959.
38. Yunker, C. E., Observations on tick tissues *in vitro*, in *Proc. 1st Int. Congr. Parasitol.*, Vol. 2, Corradetti, A., Ed., Pergamon Press, Oxford, 1966, E10, 1036.
39. Varma, M. G. R. and Pudney, M., Culture of cells from the tick, *Rhipicephalus appendiculatus* (Ixodoidea, Ixodidae), in *Proc. 3rd Int. Colloq. Invertebr. Tissue Cult.*, Rehacek, J., Blaskovic, D., and Hink, W. F., Eds., Slovak Academy of Sciences, Bratislava, 1973, 135.
40. Rehacek, J. and Brzstowski, H. W., A tick tissue culture medium based on analysis of tick hemolymph, *J. Insect. Physiol.*, 15, 431, 1969.
41. Grace, T. D. C., Establishment of four strains of cells from insect tissues grown *in vitro*, *Nature*, 195, 788, 1962.
42. Yunker, C. E., Vaughn, J. L., and Cory, J., Adaptation of an insect cell line (Grace's *Antheraea* cells) to medium free of hemolymph, *Science*, 155, 1565, 1967.
43. Vago, C. and Chastang, S., Culture *in vitro* d'un tissue nymphal de lepidoptere, *Experientia*, 14, 110, 1958.
44. Rehacek, J. and Hana, L., Notes on tick tissue cultures, *Acta Virol.*, 5, 57, 1961.
45. Mitsuhashi, J. and Maramorosch, K., Leafhopper tissue culture: embryonic, nymphal, and imaginal tissues from aseptic insects, *Contrib. Boyce Thompson Inst.*, 22, 435, 1964.
46. Medvedeva, G. I., Beskina, S. R., and Grokhovskaya, I. M., Culture of ixodid tick embryonic cells, *Med. Parasitol. Moscow*, 41, 39, 1972 (in Russian).
47. Yunker, C. E. and Cory, J., Growth of Colorado tick fever (CTF) virus in primary tissue cultures of its vector, *Dermacentor andersoni* Stiles (Acarina: Ixodidae), with notes on tick tissue culture, *Exp. Parasitol.*, 20, 267, 1967.
48. Kurtti, T. J., Munderloh, U. G., and Samish, M., Effect of medium supplements on tick cells in culture. *J. Parasitol.*, 68, 930, 1982.

Chapter 4

CHARACTERIZATION AND IDENTIFICATION OF ARTHROPOD CELL LINES

S. E. Brown and D. L. Knudson

TABLE OF CONTENTS

I. INTRODUCTION

The study of arboviruses in cell culture systems has played a role in providing the impetus for the development of invertebrate cell lines. Currently there are more than a 100 invertebrate cell lines or cell strains established.[1,2] Concomitant with the increased usage of invertebrate cell lines, there are numerous reports of the mishandling of both invertebrate and vertebrate cell lines.[3-11] Since mishandling of cell lines by mislabeling and/or contamination with extrinsic cells poses a serious problem, an approach which resolves cell line identity must be found. Before the relevant issues of cell line identification can be appropriately discussed, four points must be recognized:

1. Recognition and detection of a mishandled cell line
2. Determination of a course of action
3. Acquisition of characterization data for a cell line
4. Application of characterization data for the identification of a cell line

This chapter discusses the parameters which have been used to characterize cell lines in the past, and introduces those parameters which may prove useful to characterize and identify invertebrate cell lines in the future. These parameters are reviewed and discussed in the context of the above four basic concepts.

II. CONCEPTS

A. Recognition and Detection

The recognition and/or detection of a problem represents the first, often most difficult step in its solution. The recognition of a mishandled cell line relies on the ability to note or detect differences from its past behavior. Therefore, a certain familiarity with the cell line is often required to recognize any notable changes in the line. This familiarity may result either from previous experience in handling the cell line or from descriptive information available for the particular line. Thus, characteristics necessary to distinguish cell lines maintained in a laboratory should be established and/or recorded, and these parameters should be monitored periodically as a part of the laboratory routine.

A sudden change in the characteristics of a cell line represents the most obvious signal that it has been maltreated. However, sudden changes in cell line characteristics may also be the result of a deviation from the routine laboratory practice. Therefore a change in the characteristics can result from maltreatment of the cell line, mislabeling of the cell line, deviation from the normal laboratory routines as they relate to handling and reagents, and contamination of a cell line with extrinsic cells or other agents. If the changes are due to differences in treatment, the characteristic traits of the cell lines will return when the cell line is treated properly.

Unfortunately there are instances when change in the characteristics of a cell line will not help recognize or detect a mishandled cell line. For example, different cell lines derived from the same species may have similar characteristics which are difficult to distinguish using the current methodologies. Since extrinsic cells may represent only a small percentage of the total population of the cell line, their unique attributes may not be detected. Cell contamination involving two cell lines which exhibit similar properties may also not be recognized. In addition, a cell line which was maltreated before it was brought into the laboratory may not be recognized because the laboratory is unfamiliar with the characteristics of the cell line.

Thus, the key to recognition and detection of a mishandled cell line is familiarity with the cell lines and maintenance of accurate records. Attention must be paid to changes in the characteristics of the cell lines and the behavior. If these changes do not coincide with an expected response, a course of action must be chosen.

B. Course of Action

If the cell line was mishandled, two basic alternatives are available. One is to destroy the cell line, and the other is to attempt a determination of the identity of the line or the source of the contaminating cells. The easiest solution is to destroy the cell line and either revive an earlier passage of the line or obtain the line from a reliable source. The American Type Culture Collection (ATCC) is one example of a reliable source. The ATCC is a repository for cell lines, and it is hoped that they will expand their activities to include the increasing number of invertebrate cell lines.

On occasion, the easiest solution may not be feasible. For example, there may not be another source for the cell line. In the instance where the cell line represents a newly derived line, it must be identified. Once the decision has been made to identify the cell line, the next course of action is to characterize the cell line.

C. Characterization of a Cell Line

Characterization of a cell line relates to data acquisition on measurable biological, chemical, and physical parameters. Growth and handling characteristics, viral susceptibility, morphology, karyology, antigenic determinants, and cellular isozymes are parameters which have been used in the past to characterize invertebrate cell lines. New technologies to study additional cell-line parameters are also available. These techniques include; monoclonal antibodies, restriction enzyme analyses of mitochondrial DNA, physical mapping of mitochondrial DNA, and DNA-DNA hybridization with cell line specific probes. Therefore, the characterization of a given cell line is limited only by the technology available to measure the parameters of the line. The characterization of cell lines has progressed and will continue to progress as these technologies are applied.

D. Identification of a Cell Line

The characterization data resulting from the study of biological, chemical, and physical parameters represents the data base upon which identifications are made. Since each characterization technique has limitations, several techniques may have to be applied for a successful identification. The aim is to generate data which uniquely describes the cell line. The parameters and techniques which uniquely delineate one particular cell line from another may not be the same in another set of circumstances. Therefore the history of the cell line and a list of the other cell lines which were maintained concurrently with the mishandled cell line defines the scope of the characterization data required. Thus, careful selection of the parameters to be studied and the techniques used to measure the parameters would be prudent because it may eliminate the collection of unnecessary data.

III. BIOLOGICAL, BIOPHYSICAL, AND MOLECULAR CHARACTERIZATION

A. Growth and Handling Characteristics

Growth and handling characteristics are important in the recognition and detection of a mishandled cell line. A certain familiarity with the characteristics of each cell line develops through the routine maintenance of the cell line. Sudden changes in these distinguishing traits of a cell line can be the result of mistreatment, mislabeling, and contamination.

The ability of a cell line to grow either attached or in suspension is a notable growth characteristic. The techniques necessary to subculture a cell line can be characteristic. For example, the subculturing of an attached cell line may require one of the following: gentle agitation, scraping with a rubber policeman, or trypsinization.

Different cell lines have different growth curves. Each line has its own characteristic lag period which can range from hours to several days. The logarithmic growth period of a cell line may last 2 to 7 days. The population doubling time during logarithmic growth phase and maximal population density may be characteristic for the cell line.

Growth and handling characteristics are usually described for each cell line established. Those of a number of invertebrate cell lines have been summarized by Hink.[1] However, the definitive identification of a cell line by its growth and handling characteristics is difficult. These characteristics are easily affected by changes in the environment in which the cells are grown. Temperature, media, the age of the cells, and the handling of the cells may affect the behavior of a cell line. When a mishandling is detected by changes in growth and handling characteristics, additional parameters will need evaluation before the identity or source of contamination will be determined.

B. Viral Susceptibility

Viral susceptibility is another parameter useful in the detection of a mishandled cell line. Since arboviruses exhibit a different host range both in vivo and in vitro, the ability of a virus to grow in a particular cell line may represent a specific virus-host cell interaction. Therefore, a change in the susceptibility of a cell line to a particular virus may indicate that the cell line has been mislabeled and/or contaminated.

The virus-host interaction may result in different degrees or types of cytopathic effect (CPE). Since conditions in which the cell line is grown may alter the CPE produced, CPE, as an indicator of viral susceptibility, may not always be reliable. Fluorescent antibody staining has been a reliable alternative to CPE for the detection of viral expression in cell lines. This method provides a direct means for detecting the expression of viral proteins in cell lines.

Reviews of the host range data for arboviruses in cell culture suggest that arthropod cell lines do not necessarily develop CPE upon infection with arboviruses.[2,12] For example, viral-induced CPE was not observed in tick cell lines which have been tested. However, viral CPE has been observed in several mosquito cell lines infected with certain arboviruses. Cell death and syncytia formation are two common types of CPE that have been observed in mosquito cell lines infected with arboviruses.

Pudney et al.[12] described a situation in which virus susceptibility was an indicator of a mishandled cell line. Briefly, the host range of a number of alphaviruses was examined in mosquito cell lines, and most of the viruses tested grew in one or more lines. However, none of the viruses grew in the *Aedes vexans* or the *Culiseta inornata* cell lines. These data were inconsistent until these two cell lines were identified as the lepidopteran cell line, *Antheraea eucalypti*.[4]

Although viral susceptibility may be a useful parameter for cell-line characterization and differences in the susceptibility of a cell line to a virus may serve as an indicator of a mishandled cell line, viral susceptibility by itself may not provide the information necessary to identify a cell line.

C. Morphology

The morphologic characteristics of a number of invertebrate cell lines have been described previously.[1,13] Cell lines contain either homogeneous populations of a particular cell type or heterogeneous populations of different cell types. Size and shape are the two obvious morphologic characteristics used in cell line descriptions. Other characteristics often described include size of the nucleus in relation to the cell and absence or presence of granules and cytoplasmic extensions.

When the characteristic appearance of a cell line changes, a mishandling may be suspected. Unfortunately, cell morphology may vary as its growth conditions are altered. Thus, the morphological characteristics of a cell line may not be reliable indicators of identity.

D. Karyology

The karyotype of a chromosome set of an organism has been a useful tool in solving taxonomic problems, and karyotyping techniques have been described previously.[14] Genera are characterized by a definitive chromosome number. While there are genera in which the chromosome numbers do not vary, there are others which exhibit a wide variation in chromosome number. The karyotype, that is the number and structure of chromosomes of an organism, may exhibit a polyploid chromosome number which occurs when three or more sets of chromosomes are present. Any karyotype which has a few extra or a few less chromosomes is considered aneuploid. Two commonly recognized categories of chromosomes include the autosomes and the sex chromosomes.

Karyotype analyses are usually performed with the metaphase chromosome because the latter are easily studied exhibiting a consistent morphology, and because they are condensed. The typical metaphase chromosome consists of two chromatids attached by a centromere. The arms of the metaphase chromosomes may vary in length, width, and structural modifications, and the centromere may be located anywhere from the middle to the end of the chromosome. A symmetric karyotype is seen when the chromosomes are of the same size, and an asymmetric karyotype comprises chromosomes of different sizes. If a set of chromosomes was symmetric with centromeres located in the same place, it would be difficult to distinguish between the individual chromosomes. If the chromosomes were asymmetric with centromeres in different locations, the chromosomes may be distinguishable. A haploid set of chromosomes which are distinguishable and are arranged in decreasing order by their lengths and the positions of their centromeres represents the idiogram.

Karyological characteristics of invertebrate cell lines have been reported and the problems associated with the study of the chromosomes of insect cell lines have been discussed.[4,15-17] The chromosome number in mosquito and tick cell lines may be easily counted. While dipteran cells contain a low diploid number, the chromosome number for lepidopteran cell lines is not easy to determine because of their high ploidy. Although the lepidopteran chromosomes are asymmetric, they are too small to identify or the chromosomes have diffuse centromeres.

The karyology of a number of tick cell lines have been described.[18,19] Most *Ixodidae* cell lines have a diploid chromosome number (2n = 21 for male, 2n = 22 for female). However, a few tick cell lines are also aneuploid. In cell lines derived from hard ticks either the female or male chromosomal complement predominates. The characteristic morphologies of chromosomes from tick cell lines include autosomes which are shorter than the sex chromosomes and terminally located centromeres. Since the morphology of the chromosomes from the tick cell lines are so similar, interspecies identification of these cell lines by karyotype would be difficult.[20]

Most mosquito cell lines have a diploid chromosome number (2n = 6). Their chromosomes are small and easily counted. The differences between mosquito cell lines derived from different species are barely detectable.[21]

Karyotyping has been useful in the recognition and detection of mishandled cell lines.[4] The karyotype analysis of the mishandled *Culiseta inornata* and *Aedes vexans* cell lines demonstrated that these two cell lines were more like lepidopteran cell lines than dipteran ones. These two cell lines were eventually identified as the lepidopteran cell line, *Antheraea eucalypti.*

Although karyotype analyses of invertebrate cell lines may distinguish tick cell lines from mosquito and lepidopteran ones, interspecies identification within tick and lepidopteran cell lines may be difficult by this method. Although the interspecies identification of certain mosquito cell lines may be possible by karyotype analyses, intraspecies identification of mosquito cell lines would not be feasible.

E. Serology

Serological methods have provided techniques for the detection of antigenic determinants in the cell populations. Serological techniques have been utilized either to detect or verify the existence of a mishandled cell line. The usefulness of the immunodiffusion test, the complement-fixation test, the immunoelectrophoresis test, and the hemagglutination test for the identification of invertebrate cell lines have been evaluated.[3,4,22] Serological tests require the preparation of antigens and immune fluids for each cell line examined; the methodologies utilized in their preparation have been described previously.[23-26]

Immunodiffusion tests have been used to distinguish between invertebrate cell lines from within the orders, Lepidoptera and Diptera, of the class, Insecta.[3,4] Both the *Aedes albopictus* cell line and the *A. aegypti* cell line were distinguishable from the *Antheraea eucalypti* cell line by immunodiffusion tests.[3,4] Since the *A. albopictus* and the *A. aegypti* cell lines were cross-reactive, they were not distinguishable using the immunodiffusion test.

Aldridge and Knudson[22] evaluated complement fixation, hemagglutination, immunodiffusion, and immunoelectrophoresis for distinguishing five selected lepidopteran cell lines. The five lines represented three families with one family containing two cell lines derived from insects within the same genus. The complement-fixation tests were able to distinguish the cell lines at the family level. Goose erythrocytes did not agglutinate with the antigens under the conditions tested and thus, the hemagglutination-inhibition test was not useful. Although common cross-reactive antigen(s) were demonstrated for the cell lines tested by immunodiffusion and immunoelectrophoresis, spurs of partial identity and extra precipitin bands indicated that interfamilial differentiation was possible by immunodiffusion. Four of the five cell lines were distinguishable by immunoelectrophoresis. Since immunoelectrophoresis did allow intergeneric distinctions, the data suggested that immunoelectrophoresis was the best serologic technique.

Although serological tests have been useful in the recognition and detection of mishandled invertebrate cell lines, one of the shortcomings of the serologic approach is the requirement of reagents which include antigens and sera that exhibit specificity. Although past reports suggest a degree of specificity, the observed cross-reactivity of the antigenic determinants between the cell lines obscures its utility as an identification procedure. While the sera could be cross-adsorbed to provide greater specificity, it may not be worth the additional effort. The serological data may allow intergeneric but not interspecific distinctions.

F. Cellular Isozymes

The application of electrophoretic techniques to the study of cellular isozymes has provided a powerful tool for cell line characterization. There are numerous reports of mishandled cell lines being detected through the study of these isozymes.[3-5,7-11,27]

The usefulness of isozyme techniques for distinguishing invertebrate cell lines was initially demonstrated by Greene and Charney[3] and Greene et al.[4] Their studies indicated that cell lines derived from the tissues of insects from two different orders, Lepidoptera and Diptera, were distinguishable by their glucose-6-phosphate dehydrogenase, malate dehydrogenase, 6-phosphogluconate dehydrogenase, acid phosphatase, and lactate dehydrogenase isozyme phenotypes. The phenotype of the *Antheraea eu-*

calypti cell line was distinct from the *Aedes aegypti* and the *A. albopictus* cell lines. They used both a vertical polyacrylamide gel system[28] and a starch gel system.

The usefulness of the isozyme technique for identifying 2 dipteran and 14 lepidopteran cell lines was demonstrated by Tabachnick and Knudson.[5] The two dipteran cell lines studied were *Aedes aegypti* and *A. albopictus.* The lepidopteran cell lines studied represented eight taxonomic families with one family, Noctuidae, represented by five cell lines. The lepidopteran cell lines studied included *Estigmene acrea* (BTI-EAA), *Bombyx mori* (BM-N), *Malacosoma disstria* (IPRI-MD-108), *Lymantria disstria* (IPLB-LD-65Z), *Heliothis zea* (IMC-HZ-1; IMC-HZ-1075), *Mamestria brassicae* (IZD-MB-0503), *Spodoptera frugiperda* (IPLB-SF-21AE), *S. littoralis* (UIV-SL-573), *Trichoplusia ni* (TN-368), *Laspeyresia pomonella* (CP-1268; CP-169), *Manduca sexta* (MRRL-CH-I), and *Choristoneura fumiferana* (IPRI-CF-124). Cell extracts were prepared by a freeze-thaw method and the samples were examined using a starch gel electrophoretic method as described by Ayala et al.[29,30] The cell extracts were examined for 18 isozymes and the mobility of the isozyme bands in the gel were recorded relative to an internal standard, IPLB-SF-21AE cell line. The most predominant band of the internal standard was assigned a value of 100 for each electrophoretic system and the isozyme bands of the other cell lines were expressed relative to its mobility. The isozyme patterns from 10 of the 18 isozymes tested allowed the cell lines to be divided into groups. The isozymes tested which were not helpful in distinguishing the cell lines were: acetaldehyde oxidase (AO), adenylate kinase (ADK), alcohol dehydrogenase (ADH), alkaline phosphatase (APH), fumarase (FUM), alpha-glycerophosphate dehydrogenase (alpha-GPDH), malic dehydrogenase (MDH), and xanthine dehydrogenase (XDH). The 10 enzymes suitable for dividing the cell lines into groups were: esterase (Est), glucose-6-phosphate dehydrogenase (G6PD), hexokinase (HK), isocitrate dehydrogenase (IDH), lactic dehydrogenase (LDH), leucine amino peptidase (LAP), malic enzyme (ME), phosphoglucoisomerase (PGI), phosphoglucomutase (PGM), and tetrazolium oxidase (TO). In fact, only four isozymes, IDH, ME, PGI, and PGM, were necessary to distinguish the cell lines. In this study, the inability to distinguish TN-368 from IPLB-LD-65Z and to distinguish IPLB-HZ-1075 from IPLB-SF-21AE was not due to the resolution of the starch system, but rather the cell lines IPLB-LD-65Z and the IPLB-HZ-1075 had been mislabeled. The isozyme data, the growth and handling characteristics, and the morphology of the cells suggested that IPLB-LD-65Z was TN-368 and that IPLB-HZ-1075 was IPLB-SF-21AE. Three of the cell lines (ATC-10, ATC-15, and BM-N) showed isozyme phenotypes consistent with their species of origin.

Brown and Knudson[27] compared a simple cellulose-acetate method for the separation of cellular isozymes to the starch gel system described by Tabachnick and Knudson.[5] The 16 cell lines prepared and tested previously using the starch gel system were reexamined using the cellulose-acetate system as described by Kreutzer et al.[31] and Kreutzer and Christensen.[32] The cell extracts were tested for their PGM, PGI, ME, and IDH activities. The mobilities of the bands relative to the standard cell line, IPLB-SF-21AE, were calculated and recorded. The data generated by the cellulose-acetate and the starch gel electrophoretic systems were comparable except for one cell line. Using the cellulose-acetate system, IZD-MB-0503 was identical to IPLB-LD-65Z and TN-368. This difference in the results between the two systems was related to the better resolution of the starch gel system.

Brown and Knudson[6] extended the number of cell lines characterized for their isozyme phenotypes using the previously described cellulose-acetate technique to 5 mosquito species, 12 lepidopteran species, and 1 tick species. The mosquito cell lines studied were *Aedes aegypti* (ATC-10), *A. albopictus* (ATC-15), *A. pseudoscutellaris* (LSTM-AP-61), *Anopheles stephensi* (LSTM-AS-43), and *Toxorhynchites amboinen-*

sis (PRU-TA-9 and PRU-TA-42). The tick cell line studied was *Rhipicephalus appendiculatus* (LSTM-RA-243). The cell extracts were prepared similarly, and the isozyme phenotypes studied were PGM, PGI, ME, and IDH. An *Aedes aegypti* cell line designated as ATC-10* which was received during this study was identified as *A. albopictus.* The method was able to distinguish the cell lines derived from three different species of *Aedes* mosquitos. The technique did not distinguish between the different cell lines derived from either *A. aegypti, A. albopictus,* or *T. amboinensis.* Therefore, the cellulose-acetate system as described allowed interspecies but not intraspecies distinction.

Herrera and Mukherjee[7] reported the separation of cellular isozymes using slab acrylamide electrophoresis system as described by Davis.[33] In this study cell extracts from eight insect cell lines were prepared by a homogenization technique. The mosquito cell lines studied were *A. aegypti* (ATC-10), *A. aegypti* (Mos. 20A), *A. albopictus* (ATC-15), *Anopheles gambiae* (Mos. 55), *A. stephensi* (Mos. 43), and *Culex quinquefasciatus.* The additional two cell lines were *Trichoplusia ni* and *Drosophila melanogaster.* These lines were tested for their isozyme phenotypes using 15 dehydrogenases. The 15 enzymes tested were: ADH, galactose 6-phsophate dehydrogenase (Gal-6PDH); G6PD, glutamate dehydrogenase (GDH); glyceraldehyde 3-phosphate dehydrogenase (G3PD); alpha-GPDH, D-3-hydroxybutyrate dehydrogenase (HBDH); IDH, LDH, MDH, octanol dehydrogenase (ODH); 6-phosphogluconate dehydrogenase (6PGD); retinol dehydrogenase (RDH); sorbitol dehydrogenase (SDH); and xanthine dehydrogenase (XDH). The gels were stained according to procedures described by Shaw and Prasad,[34] and the relative mobilities were calculated by the distance moved by the isozyme divided by the distance traveled by the bromophenol blue marker dye. The acrylamide system verified the cellular contamination of the Mos. 55 line used in this study. Since the isozyme phenotypes for the two particular cell lines ATC-10 and Mos. 20A were different, then this method may allow intraspecies distinctions.

The Authentikit® system developed by Corning (Corning Glass Works, Corning, N.Y.) for the separation of cellular isozymes may be by far the easiest and most convenient method. This system consists of three individual parts: equipment, reagents, and an interpretation system. The equipment includes sample applicators, electrophoresis cells, power supplies, and an oven/incubator. The reagent system consists of preformed gels, buffer, enzymatic substrates, standards, controls, extraction buffer, and stabilization buffers. The interpretation system includes the information needed to analyze the data. The kit eliminates the hassle involved in making the buffers and staining reactions and includes a stabilization buffer which allows long-term storage of prepared cell extracts. Thus, each cell line carried in a laboratory can have its own stored reference. If contamination or mislabeling is suspected, the cell line can be tested against itself. The procedures include a protocol check to insure that the samples have sufficient enzyme activities for the test and that the enzyme activity for the samples falls into an appropriate activity range. The system has been used to characterize two new cell lines developed from the beetle *Diarotica undecimpunctata.*[35] In addition, the Authentikit® system has been compared with the starch and cellulose-acetate systems in this laboratory.[5,6,27] The cell lines prepared and tested previously by Tabachnick and Knudson[5] and Brown and Knudson[6] which were available were retested using the Authentikit® system. Cell extracts were prepared by an osmotic shock/shear procedure. The isozymes tested using this system were nucleotide phosphatease (NP), glucose-6-phosphate dehydrogenase (G6PD), malate dehydrogenase (MDH), mannose phosphate isomerase (MPI), peptidase b (PepB), aspartate aminotransferase (AST), lactate dehydrogenase (LDH), and malic enzyme (ME). These enzymes also distinguished the cell lines described by Tabachnick and Knudson[5] and Brown and Knudson.[6,27] Additionally, five *Leishmania* spp. were tested, *Leishmania B. panamensis* (WR-120), *L. chagasi* (WR-121B), *L. donovani* (LCR-L52), *L. M. amazonensis* (WR-303), *L. M. mexicana* (L-11), and *L. tropica major* (WR-309) were also distinguished.

Examination of cellular isozyme phenotypes is a useful technique for the recognition and detection of mishandled cell lines. Isozyme data have identified mishandled cell lines which exhibit defined isozyme phenotypes. Since isozyme analyses are being reported for new cell lines and there is more literature available concerning isozyme phenotypes for established cell lines, investigators will be able to compare their data to well-documented standards. In addition, reports on isozyme phenotypes for invertebrate cell lines are increasing and the isozymes which demonstrate the greatest variability in invertebrate cell lines are being recognized. In the future only the isozymes which demonstrate the greatest diversity in invertebrate cells may require testing.

IV. NEW TECHNOLOGIES FOR CELL LINE IDENTIFICATION

A. Cellular DNA Characterization

The parameters used to identify invertebrate cell lines are the result of the cellular gene expression. Technology is now available which permits a more direct evaluation of the cellular DNA. Eukaryotic cellular DNA is found in the mitochondria and the nucleus of cells.

1. Mitochondrial DNA (mDNA)

The mDNA is easily distinguished from the nuclear DNA (nDNA) of eukaryotic cells.[36] The mDNA is smaller (approximately 18 Kbp), it may exhibit a different base pair composition than the nDNA, and mDNA is a double stranded, circular, and supercoiled molecule. Although mDNA represents approximately 1% of the total DNA content in cells, it is more amenable to taxonometric studies than nDNA because it is not as complex. Sequence variations may occur in mDNA between species and between mDNA from within the same species. Since mDNA is maternally inherited, a lineage which can be followed is established.[37] Thus, these properties may make mDNA an ideally suited nucleic acid sequence to examine for its potential in cell line characterization.

The sequence variation in mDNA can be measured either directly by nucleic acid sequencing techniques or indirectly by restriction enzyme analysis and nucleic acid hybridization techniques. Restriction endonucleases which recognize specific short DNA sequences cleave a phosphodiester bond in each DNA strand at or within the recognition sequence. When the digested DNA is electrophoresed through an agarose or acrylamide gel, a specific profile of fragments separated by their size will be evident. These DNA fragments may be visualized in the gels by staining with ethidium bromide or by tagging the DNA with radioisotopes and using autoradiographic techniques.

The mDNA may be purified[38] from cell lines and cleaved separately with a number of different restriction enzymes. Mitochondrial DNAs from a variety of different cell lines will undoubtedly produce distinctive restriction profiles. Even when different mDNAs may have the same number of cleavage sites for an enzyme, the molecular weight values of the fragments may be different. When mDNAs may have a different number of recognition sites for the restriction enzymes examined the fragment profiles will have a different number of bands. The ordering and positioning of the restriction sites for the construction of a physical map of the mDNA may be required to distinguish mDNAs which have similar restriction enzyme profiles. Techniques for restriction enzyme analysis and the construction of a physical map have been described previously.[39]

The *Eco*RI and *Hind*III restriction enzyme profiles for mDNA for the cell lines

Trichoplusia ni and *Spodoptera frugiperda* yield distinguishable fragment profiles in agarose gels (unpublished data). The size of the mDNA from a *Drosophila* cell line was 18 Kbp and its was restricted into three fragments by the restriction enzyme *Hae*III. The *Hae*III restriction enzyme profiles of Schneider's *Drosophila* line 2 and line K were distinguishable.[40]

The mDNA data available suggest that restriction enzyme digestion of mDNA will be an appropriate technique for both interspecies and intraspecies identification of invertebrate cell lines. If restriction enzyme profiles of the mDNA are not sufficient to characterize two closely related cell lines, then either the construction of a physical map or DNA hybridization experiments may be necessary to distinguish two closely related mDNAs.

An alternative strategy would be to insert the restricted mDNA fragments into a bacterial cloning vector such as the pUC series derivatives of pBR322.[39,41] These recombinant clones or mDNA libraries may prove extremely useful as potential cell line specific probes in "dot-blot"[42] hybridizations with the smaller fragments having the greater probability to be cell line specific. If a cell line is thought to be mislabeled or contaminated, the cells from that line could be dotted onto nitrocellulose membranes and the cellular DNA could be liberated. Recombinants containing the cell-specific DNAs necessary to identify the cell line under scrutiny would be purified, labeled, and hybridized individually to the dot blots. Instead of a radioactive label, the recombinant probe may be nick-translated with a biotinylated dUTP which would be detected using a streptavidin-horseradish detection system.[43] Since the biotin label is stable, the probes could be used over an indifinite period of time.

2. Nuclear DNA (nDNA)

The measurement of nDNA by restriction enzyme analyses may not be as rewarding because of the complexity of the molecules. The high molecular weight nDNA would be purified and restricted by enzymes to produce a range of fragments. A selection of approximately 10 Kbp fragments would be made by electrophoresis and extraction from an appropriate region of the gel. Genomic libraries could be constructed in bacteriophage lambda vectors by insertion of the isolated cellular fragments. The genomic libraries could be screened for recombinants containing cellular DNA sequences (see discussion above).

Although the potential of these approaches for the characterization and identification of invertebrate cell lines is clear, it has not been fully exploited. The establishment of genomic and mitochondrial libraries containing cell line specific probes will be invaluable to cell line identification in the future. One may envisage recombinant probes with various levels of taxonomic specificity.

B. Monoclonal Antibodies

Monoclonal antibodies represent an attractive alternative to the heterologous antisera which has been used in the past and it may find applications in cell line identification. Advantages and disadvantages in the use of monoclonal antibodies have been described by Kennett.[44] The advantages which may be helpful to cell line identification include: monoclonal antibodies react with a single antigenic determinant, results from using monoclonal antibodies are reproducible, monoclonal antibodies may be produced in large quantities, and monoclonal antibodies will detect components in a mixture. A disadvantage is that monoclonal antibodies may not be used in either immunodiffusion or immunoelectrophoresis techniques because the monoclonal antibody does not precipitate in gels. The methodology involved in producing monoclonal antibodies has been described in detail elsewhere.[45]

Monoclonal antibodies are currently in use in many areas of research. For example, these antibodies have been used to differentiate subsets of T cells.[44] Hence, they may allow interspecies and intraspecies distinctions for invertebrate cell lines. However, an investigator interested only in identifying an occasional cell line may not wish to use these techniques unless the monoclonal antibody which can distinguish the cell line of interest is already available. Therefore, the future establishment of a bank of monoclonal antibodies against invertebrate cell lines will be invaluable to cell line identification.

V. SUMMARY AND CONCLUSIONS

The concepts relevant to cell line identification have been described and discussed. The characteristics necessary to distinguish the cell lines which are maintained in a laboratory should be determined before a problem develops. This precaution may make the characterization and therefore identification process easier. The recognition of a mishandled cell line requires a certain familiarity with the characteristics of the cell line. Once mishandling has been recognized, the easiest course of action is to destroy the cell line. If this is not feasible, then the identification of the cell line through characterization is necessary.

Accurate record keeping is a prerequisite. The records should document the cell lines and provide a laboratory history of other lines being maintained concurrently at the time of the mishandling. Documentation for the cell line should include growth and handling characteristics, karyotype, and known isozyme phenotypes.

In conclusion, the problem of cell line identification becomes a "quality control" exercise. When mishandling occurs and is detected and when the autoclave is not an acceptable solution, the technique in widespread use with the greatest sensitivity is isozyme analysis. The Authentikit® system may represent the easiest method to use for the occasional or routine identifications. The ability to store prepared cell extracts for an indefinite period of time is an attractive feature. Thus, cell extracts may be prepared for all cell lines which are maintained in a laboratory and these extracts may be stored for reference. If mishandling is suspected, the cell line can be tested against itself using the reference reagent.

The new recombinant DNA and monoclonal antibody techniques for cell line characterization have great potential for cell line identification. If cell lines are not distinguishable by the more conventional methods, then the new technologies offer additional options. The production of genomic libraries of cell line-specific probes and banks of monoclonal antibodies against cell lines warrants continued support.

REFERENCES

1. **Hink, W. F.,** A compilation of invertebrate cell lines and culture media, in *Invertebrate Tissue Culture: Research Applications,* Maramorosch, K., Ed., Academic Press, New York, 1976, 319.
2. **Knudson, D. L. and Buckley, S. M.,** Invertebrate cell culture methods for the study of invertebrate-associated animal viruses, in *Methods in Virology,* Maramorosch, K. and Koprowski, H., Eds., Academic Press, New York, 1977, 323.
3. **Greene, A. E. and Charney, J.,** Characterization and identification of insect cell cultures, in *Current Topics in Microbiology and Immunology,* Vol. 55, Weiss, E., Ed., Springer-Verlag, New York, 1971, 51.
4. **Greene, A. E., Charney, J., Nichols, W. W., and Coriell, L. L.,** Species identity of insect cell lines, *In Vitro,* 7, 313, 1972.
5. **Tabachnick, W. J. and Knudson, D. L.,** Characterization of invertebrate cell lines. II. Isozyme analyses employing starch gel electrophoresis, *In Vitro,* 16, 392, 1980.

6. Brown, S. E. and Knudson, D. L., Characterization of invertebrate cell lines. IV. Isozyme analyses of dipteran and acarine cell lines, *In Vitro*, 18, 347, 1982.
7. Herrera, R. J. and Mukherjee, A. S., Electrophoretic characterization and comparison of dehydrogenases from eight permanent insect cell lines, *Comp. Biochem. Physiol.*, 72, 359, 1982.
8. Petersen, W. D., Jr., Simpson, W. F., and Hukku, B., Cell culture characterization: monitoring for cell identification, in *Methods in Enzymology*, Vol. 57, Academic Press, New York, 1979, 164.
9. Harris, N. L., Gang, D. L., Quay, S. C., Poppema, S., Zamecnik, P. C., Nelson-Rees, W. A., and O'Brien, S. J., Contamination of Hodgkin's disease cell cultures, *Nature*, 289, 228, 1981.
10. Nelson-Rees, W. A., Daniels, D. W., and Flandermeyer, R. R., Cross-contamination of cells in culture, *Science*, 212, 446, 1981.
11. O'Brien, S. J., Shannon, J. E., and Gail, M. H., A molecular approach to the identification and individualization of human and animal cells in culture: isozyme and allozyme genetic signatures, *In Vitro*, 16, 119, 1981.
12. Pudney, M., Leake, C. J., and Buckley, S. M., Replication of arboviruses in arthropod *in vitro* systems: an overview, in *Invertebrate Cell Culture Applications*, Maramorosch, K. and Mitsuhashi, J., Eds., Academic Press, New York, 1982, 159.
13. Hink, W. F., Insect tissue culture, *Adv. Appl. Microbiol.*, 15, 157, 1972.
14. Sharma, A. K. and Sharma, A., *Chromosome Techniques: Theory and Practice*, Butterworths, Boston, 1980, 711 pp.
15. Hink, W. F. and Ellis, B. J., Establishment and characterization of two new cell lines (CP-1268 and CP-169) from the codling moth, *Carpocapsa pomonella* (with a review of culture of cells and tissues from Lepidoptera), in *Current Topics in Microbiology and Immunology*, Vol. 55, Weiss, E., Ed., Springer-Verlag, New York, 1971, 19.
16. Nichols, W. W., Bradt, C., and Bowne, W., Cytogenetic studies in culture from the class Insecta, in *Current Topics in Microbiology and Immunology*, Vol. 55, Weiss, E., Ed., Springer-Verlag, New York, 1971, 51.
17. Schneider, I., Karyology of cells in culture: characteristics of insect cells, in *Tissue Culture: Methods and Applications*, Kruse, P. K., Jr. and Patterson, M. K., Jr., Eds., Academic Press, New York, 1973, 788.
18. Yunker, C., Cory, J., and Meibos, H., Continuous cell lines from embryonic tissues of ticks (Acari: Ixodidae), *In Vitro*, 17, 139, 1981.
19. Kurtti, T. J. and Munderloh, U. G., Tick cell culture: characteristics, growth requirements, and applications to parasitology, in *Invertebrate Cell Culture Applications*, Maramorosch, K. and Mitsuhashi, J., Eds., Academic Press, New York, 1982, 195.
20. Oliver, J. H., Jr., Cytogenetics of ticks and mites, *Annu. Rev. Entomol.*, 22, 407, 1977.
21. Hirumi, H., Viral, microbial, and extrinsic cell contamination of insect cell cultures, in *Invertebrate Tissue Culture: Research Applicatoins*, Maramorosch, K., Ed., Academic Press, New York, 1976, 233.
22. Aldridge, C. A. and Knudson, D. L., Characterization of invertebrate cell lines. I. Serologic studies of selected lepidopteran lines, *In Vitro*, 16, 384, 1980.
23. Clarke, D. H. and Casals, J., Techniques for hemagglutination and hemagglutination-inhibition with arthropod-borne viruses, *Am. J. Trop. Med. Hyg.*, 7, 561, 1958.
24. Brandt, W. E., Buescher, E. L., and Hetrick, F. M., Production and characterization of arbovirus antibody in mouse ascitic fluid, *Am. J. Trop. Med. Hyg.*, 16, 339, 1967.
25. Casals, J., Immunological techniques for animal viruses, in *Methods in Virology*, Vol. 3, Maramorosch, K. and Koprowski, H., Eds., Academic Press, New York, 1967, 113.
26. Curtain, C. C., Baumgarten, A., and Pye, J., Co-precipitation of some cryomacroglobulin with immunoglobulins and their fragments, *Arch. Biochem. Biophys.*, 112, 37, 1965.
27. Brown, S. E. and Knudson, D. L., Characterization of invertebrate cell lines. III. Isozyme analyses employing cellulose-acetate electrophoresis, *In Vitro*, 16, 829, 1980.
28. Raymond, S., Acrylamide gel electrophoresis, *Ann. N.Y. Acad. Sci.*, 121, 350, 1964.
29. Ayala, F. J., Powell, J. R., Tracey, M. L., Mourao, C. A., and Perez-Salas, S., Enzyme variation in the *Drosophila willistoni* group. IV. Genic variation in natural populations of *Drosophila willistoni*, *Genetics*, 70, 113, 1972.
30. Ayala, F. J., Powell, J. R., Tracey, M. L., Mourao, C. A., and Perez-Salas, S., Genic variation in natural populations of five *Drosophila* species and the hypothesis of the selective neutrality of protein polymorphisms, *Genetics*, 77, 343, 1974.
31. Kreutzer, R. D., Posey, F. T., and Brown, P. A., A fast and sensitive procedure for identifying genetic variants of phosphoglucomutase in certain genera of mosquitos, *Mosq. News*, 37, 407, 1977.
32. Kreutzer, R. D. and Christensen, H. A., Characterization of Leishmania spp. by isozyme electrophoresis, *Am. J. Trop. Med. Hyg.*, 29, 199, 1980.
33. Davis, B. J., Disc electrophoresis. II. Method and application to human serum proteins, *Ann. N.Y. Acad. Sci.*, 121, 404, 1964.

34. **Shaw, C. R. and Prasad, R.,** Starch gel electrophoresis of enzymes — a compilation of recipes, *Biochem. Genet.,* 4, 297, 1970.
35. **Lynn, D. E. and Stoppleworth, A.,** Established cell lines from the beetle, *Diabrotica undecimpunctata* (Coleoptera:Chrysomelidae), *In Vitro,* 20, 365, 1984.
36. **Lehninger, A. L.,** *Biochemistry,* Worth Publishers, New York, 1978, 1104 pp.
37. **Brown, W. M.,** Evolution of animal mitochondrial DNA, in *Evolution of Genes and Proteins,* Nei, M. and Koehn, R. K., Eds. Sinauer Assoc., Sunderland, Mass., 1983, 62.
38. **Fouts, D. L., Manning, J. E., and Wolstenholme, D. R.,** Physiochemical properties of kinetoplast DNA from *Crithidia acanthocephali, Crithidia luciliae,* and *Trypanosoma lewisi, J. Cell Biol.,* 67, 378, 1975.
39. **Maniatis, T., Fritsch, E. F., and Sambrook, J.,** *Molecular Cloning. A Laboratory Manual,* Cold Spring Harbor Laboratory, New York, 1982, 545.
40. **Manteuil, S., Hamer, D. H., and Thomas, C. A., Jr.,** Regular arrangement of restriction sites in *Drosophila* DNA, *Cell,* 5, 413, 1975.
41. **Messing, J., Crea, R., and Seeburg, P. H.,** A system for shotgun DNA sequencing, *Nucleic Acids Res.,* 9, 309, 1981.
42. **Kafatos, F., Jones, W. C., and Efstratiadis, A.,** Determination of nucleic acid sequence homology and relative concentrations by a dot blot hybridization procedure, *Nucleic Acids Res.,* 7, 1541, 1979.
43. **Langer, P. R., Waldrop, A. A., and Ward, D. C.,** Enzymatic synthesis of biotin-labeled polynucleotides: novel nucleic acid affinity probes, *Proc. Natl. Acad. Sci.,* 78, 6633, 1981.
44. **Kennett, R. H.,** Hybridomas: a new dimension in biological analyses, *In Vitro,* 17, 1036, 1981.
45. **Oi, V. T. and Herzenberg, L. A.,** Immunoglobulin-producing hybrid cell lines, in *Selected Methods in Cellular Immunology,* Mishell, B. and Shiigi, S., Eds., W. H. Freeman, San Francisco, 1980, 351.

Chapter 5

IN VITRO IMMUNOENZYMATIC DETECTION AND SCREENING OF ARTHROPOD-BORNE TOGAVIRUS ANTIGENS AND ANTIBODIES

E.Kurstak, P. Tijssen, and C. Kurstak

TABLE OF CONTENTS

I. INTRODUCTION

The Togaviridae includes four genera (alphaviruses, flaviviruses, rubiviruses, and pestiviruses).[29] Only *Alphavirus* and *Flavivirus* multiply both in arthropods and vertebrates and have an arthropod-vertebrate-arthropod transmission cycle. The structure, composition, and replication of these two genera have been extensively studied.[1,18,37] Though these viruses infect both vertebrate and invertebrate cells in tissue culture, there is an essential difference. Host cell macromolecular synthesis in vertebrate cells is inhibited in contrast to the noncytolytic infection of permissive invertebrate cells, though some exceptions have been reported.[39,46]

In vitro studies and diagnosis of alpha- and flaviviruses by serological methods are often hampered by the close antigenic resemblance among these viruses. On the other hand, clinical cases reach the physician often when the viraemia has decreased significantly and direct identification becomes difficult. Classical methods such as hemagglutination-inhibition and complement-fixation are fairly rapid but not very specific or sensitive. Virus neutralization is a lengthy and costly procedure. New developments in enzyme immunoassays have enabled rapid screening and detection of *Alpha-* or *Flavivirus* antigens and their corresponding antibodies.

II. COMPARATIVE ASPECTS OF ARTHROPOD-BORNE TOGAVIRUSES

The type species of the *Alphavirus* (Arbovirus antigenic group A) genus is the Sindbis virus. Viruses belonging to this genus have about 6% genome nucleic acid (RNA) by weight of particle (70×10^6 daltons), i.e., a single stranded RNA of about 4.3×10^6 daltons and 42S of (+) polarity. About 60% (by weight) of the virus particle consists of proteins, i.e., a nonglycosylated capsid (C) protein of about 30 kDa and two envelope glycoproteins (E1 and E2) of about 50 kDa present in the membrane as spikes. In the case of the Semliki Forest virus, a third viral glycoprotein (E3) is associated with the spikes (the corresponding protein of other alphaviruses is released into the culture fluid). The spherical, enveloped 70 nm-particles have a density in sucrose of about 1.21 and a sedimentation coefficient of 280S. They replicate in the cytoplasm and mature by the budding of preassembled nucleocapsids through the plasma membrane. During the replication a complete complementary transcript ("−") of the genome is produced which serves as a template for the 42S and 26S mRNAs. Both mRNAs are capped and polyadenylated.[23] The 42S mRNA codes for nonstructural proteins,[40] whereas the subgenomic 26S mRNA which is produced in threefold molar excess codes for structural proteins. Among the 25 species of alphaviruses the best known are Sindbis, Semliki Forest, Chikungunya, Eastern, Western, and Venezuela equine encephalomyelitis, O'Nyong-Nyong, and Ross River.

The type species of the *Flavivirus* (Arbovirus group B) is the yellow fever virus. The flaviviruses are considerably smaller than the alphaviruses (i.e., only 50 nm diameter). The 40S (4.2×10^6 daltons) genomic RNA is associated with a 13 kDa nucleocapsid (C) protein. The envelope contains one glycoprotein (E, 55 kDa) and one protein (M 8 kDa). The viral RNA like the RNA of alphaviruses is capped, but lacks the 3'- terminal poly(A) sequence.[47] Deproteinized RNA is infectious and evidence for subgenomic RNA is lacking. Some flaviviruses multiply in mosquitoes (e.g., dengue types, Kunjin, Japanese encephalitis, Murray Valley encephalitis, West Nile, St. Louis encephalitis) others in ticks (Omsk hemorrhagic fever, tick-borne encephalitis), whereas for others the vector is unknown. In contrast to alphaviruses, the flaviviruses presumably mature by budding through intracytoplasmic membranes (endoplasmic reticulum). Brinton[6] noted that resistant cultures enhanced amplification of defective interfering particles.

The replication of flaviviruses in cytoplasm is characterized by a latent period of about 12 to 16 hr. Transcription from the plus-strand 44S RNA genome occurs in the perinuclear region. The very high RNase resistance of the 20S RNA containing the templates indicates only one nascent strand. This strand and the RNA polymerase are released slowly upon completion of transcription. Translation of viral proteins from the 44S plus-strand RNA takes place in rough endoplasmic reticulum.[48] Protein synthesis increases from about 12 hr to a maximum at about 24 hr after infection.[42] The control of translation seems to be strongly influenced by the K^+ ion concentration.[31]

III. ANTIGENIC PROPERTIES OF ARTHROPOD-BORNE TOGAVIRUSES

Alphaviruses all seem to be part of the Arbovirus antigenic group A.[9] Core proteins are widely cross-reactive and are used for the group demarcation, whereas envelope proteins are subgroup (complex) or type specific (Eastern equine encephalitis, Semliki, Venezuelan equine encephalitis, Western equine encephalitis). Though ideally both the virus is isolated and antibodies to the virus detected, only the latter is usually possible. The antigenic relationships between the various alphaviruses necessitate stringent assay conditions. Common serological tests such as hemagglutination-inhibition, complement-fixation, and virus neutralization methods should be performed with different antigens, though situations may exist in which only a given virus of the taxon is known to be active in that geographical area. Complement-fixation is more specific than the hemagglutination test, but less specific than the neutralization test. The neutralization tests are, however, reagent consumptive, expensive, and labor intensive. Increasingly, immunofluorescence and more recently immunoenzymatic techniques are used for identification of this group of viruses due to significant advantages.[26] The approximately 55 flaviviruses were assembled in their taxon primarily on the basis of serological cross-reactivity (Arbovirus, antigenic group B).[9,17] Recently some viruses have been described which are antigenically unrelated but share other properties of the flaviviruses.

There are at least three different *Flavivirus* antigens, core protein, envelope glycoprotein, and the four or five distinct nonstructural antigens. Among the classical serological methods to diagnose flaviviruses, hemagglutination, complement-fixation, and neutralization, hemagglutination yields the strongest cross-reactions and neutralization yields the least cross-reactivity. The degree of cross-reactivity allows the distinction of subgroups or complexes which cross-react strongest among themselves. About half of the flaviviruses can be distributed in three complexes (i.e., the dengue types; the tick-borne complex; the St. Louis encephalitis — Japanese encephalitis — West Nile virus complex).

Though clinical and epidemiological data may suggest the involvement of a certain virus, only laboratory procedures may attain a definite virus identification. These procedures can be based on direct virus isolation and detection or by the identification of the antibodies developed. Blood is the specimen of choice for the direct isolation of virus. However, patients generally see the physician when signs and symptoms point to severe forms of illness, at which stage the viremia decreased appreciably. If arthropod-borne togavirus involvement is suspected, it is necessary to test for antibodies.

Either newborn mice or vertebrate cell cultures (VERO, BHK-21, HeLa, LLC-MK$_2$ cells) are inoculated with the specimen for virus isolation. The use of arthropod cells (*Aedes albopictus*) requires demonstration of the virus either by serological methods or, in some cases, the subsequent use of host cells which show a CPE. It is essential to exclude other microorganisms such as indigenous viruses in laboratory animals or cell cultures, e.g., by testing the isolate against polyvalent Arbovirus sera such as antigenic groups A, B, C, Bunyamwera, Bwamba, VSV, and other groups[12] which are available from the Research Resources Branch, National Institute of Allergy and Infectious Diseases, Bethesda, Md., 20205.

Blood or cerebrospinal fluid collected is refrigerated at 4°C (heparinized, 0.16 mg/mℓ, whole blood is preferable for some viruses such as Colorado tick fever) or at −60°C if immediate inoculation is not possible. For antibody studies it is also necessary to collect after 3 weeks convalescent-phase sera. Most arboviruses cause illness in newborn mice. Litters of 1- to 4-day old Swiss mice (as free as possible from contaminating viruses) are inoculated intracerebrally with 0.015 mℓ of the specimen. Alternatively, cell cultures can be inoculated with 0.1 mℓ of the specimen and checked for CPE[45] or plaques (under agar overlay).[32] If dengue viruses are suspected, mosquitoes may be inoculated and after 7 days screened for the presence of an antigen in their heads by immunohistochemical procedures.[33]

IV. PREPARATION OF REAGENTS FOR IMMUNOENZYMATIC TESTS

Immune sera for immunoenzymatic techniques are preferably prepared in mice because they are susceptible to most arboviruses but do not usually produce mouse tissue antibody. Mice are immunized with uncentrifuged infected, 10% mouse brain suspension in saline by intraperitoneal injections (0.2 mℓ on days 2 and 30) and are bled on day 37 (or on day 1, 3, 10, 15, and 20 with bleeding on day 27).[36] Freund's complete adjuvant may increase the titer.[5] For polyvalent or grouping sera the various viruses are mixed. In cases where one of the viruses is lethal when injected intraperitoneally, it may be inactivated by incubation of the 10% brain suspension with 0.1% β-propiolactone in saline for 1 hr at 37°C.

The suitability of mice for the production of antibodies makes the production of monoclonal antibodies a logical choice to obtain specific reagents and to avoid cross-reactivity among otherwise serologically-related viruses. The production, purification, and characterization of monoclonal antibodies against togavirus antigens has been described in detail elsewhere.[3,4] However, it is important to note that the use of different monoclonal antibodies directed to the same antigen but not competing for the same epitope enhances antigen detection by the immunoperoxidase method.[43] This is due to a gain in avidity (affinity bonus) since in the constantly rapid association-dissociating process of the antigen-antibody complex the antigen and antibodies become less physically separated if two or more antibodies keep the antigen in place.[27,49,50]

The immunoglobulin fraction from sera can be purified by a plethora of methods[28] such as salt or organic solvent precipitation, chromatography, electrophoresis, isoelectric focusing, isotachophoresis, ion-exchange chromatography, and affinity chromatography.

The approximate purity and quantity of IgG are most constantly established by the measurement of the optical densities of the protein sample at 251 and 278 nm. The absorbance ratio at 278/251 nm should be 2.5 and the absorbance at 278 nm (1 cm optical pathway) divided by 1.35 yields the concentration (in mg/mℓ).

Peroxidase can be purified by a recently described simple method.[27,41] Peroxidase is dissolved in sodium phosphate buffer, 2.5 mM, pH 8.0 at about 5 mg/mℓ. A DEAE-Sepharose column is equilibrated with the same buffer. The enzyme sample (about 5 mg protein per mℓ gel) is applied. The impurities are retained, whereas the most active isozyme of peroxidase (C) passes directly (a brown fraction). The RZ (i.e., the optical density ratio at 401 and 275 nm, reflecting the hemin content of the protein) will be better than any of the commercial offered peroxidases (i.e., higher than 3.0) and the specific activity of the enzyme is higher. Peroxidase at 1 mg/mℓ has an optical density at 403 nm (1 cm optical path length) of 2.25 cm²/mg.

Direct chemical linkage of peroxidase to immunoglobulin is most popular. This method is based on the fact that the peroxidase protein is located in a shell of carbohydrates. The carbohydrates can be oxidized gently so that the enzyme activity is hardly affected. This oxidation results in the transformation of vicinal glycol groups into aldehyde groups. These aldehyde groups may form Schiff's bases with amino groups of the IgG. These unstable bases can be stabilized by reduction with sodium borohydride.

In detail 5 mg of purified peroxidase is dissolved in 0.5 mℓ of freshly prepared 0.1 M sodium bicarbonate (prepared in distilled water) in a small tube. Subsequently, 0.5 mℓ of 10 mM sodium metaperiodate is added, the tube closed, and incubated in the dark for 2 hr at 20°C. IgG (15 mg/mℓ) is equilibrated in 0.1 M sodium carbonate buffer, pH 9.2, and added to the peroxidase solution after the activation of the enzyme followed by the addition to a small column containing an amount of Sephadex G-25 equaling one sixth of the combined weight of the peroxidase and IgG samples. After another incubation for 3 hr at room temperature, the conjugate is eluted from the Sephadex by chasing with 10^{-4} M sodium hydroxide. Freshly prepared sodium borohydride (5 mg/mℓ in 10^{-4} M NaOH) is added (1/20 volume) followed after half an hour by a 3/20 volume of again freshly prepared sodium borohydride. After an incubation of 1 hr, the preparation may be purified. The method used in our laboratory is convenient and effective. The conjugate solution is mixed with an equal volume of saturated ammonium sulfate and equilibrated for 30 min to 1 hr. This results in the precipitation of IgG and IgG conjugates whereas peroxidase or albumin-peroxidase conjugates remain soluble. Low-speed centrifugation (10 min at 5000 xg) allows the recovery of the IgG and IgG-peroxidase in the pellet. Free IgG is then removed with an affinity chromatography step on Con A-Sepharose (equilibrated with CA buffer: 0.1 M acetate buffer pH 6.0, supplemented with 1 M sodium chloride, 1 mM calcium chloride, 1 mM magnesium chloride, and 1 mM manganese chloride). The pellet is resolubilized in CA buffer and is passed on the Con A-Sepharose column. Con A has an affinity for peroxidase but not for IgG so that free IgG passes freely. Pure peroxidase conjugate can then be described by CA buffer supplemented with 0.1 M α-methyl-D-mannopyranoside which competes with peroxidase for the same sites on Con A but is added in excess. Storage of the pure conjugate is most convenient by adding 5 mg BSA per mℓ and an equal volume of glycerol and storing at −20°C (remains liquid so that small aliquots may be retrieved whenever necessary). Filtration of the conjugate is advised since peroxidase is very sensitive to both bacteria and bacteriostatic agents.

An increasing popular alternative is the biotinylation of both enzyme and IgG and the linking of these two moieties by avidin or streptavidin. Protein concentration should be 10 to 20 mg/mℓ and the proteins should be equilibrated with 0.2 M sodium bicarbonate. Biotinyl-N-hydroxysuccinimide in dimethyl sulfoxide at about 0.1 M. The molar ratio (per amino group) in which BNHS is added to IgG or peroxidase should be 2 and 10, respectively, (1gG 50 to 70 amino groups per molecule; peroxidase about 2 amino groups per molecule). The mixture is incubated for a few hours with intermittent stirring. Unreacted biotin is removed by extensive dialysis. Biotinylated enzyme can be completed just before use with avidin by adding a fourfold excess of the latter.

V. ENZYME IMMUNOHISTOCHEMISTRY OF ARTHROPOD-BORNE TOGAVIRUSES

Immunohistochemical techniques have recently been rapidly increasing in popularity although fluorescent-antibody tests were described almost 2 decades ago.[16] This is due to rapid advances in the immunoenzyme assay technology, to the increasing availability of monoclonal antibodies (consult *Linscott's Directory of Immunological and Biolog-*

ical Reagents, 40 Glenn Drive, Mill Valley, Calif., 94941) and to the inherent advantages of immunoenzyme techniques. The outline of these techniques has been reviewed in extenso[24,25,27,28,50] with special emphasis on arthropod-borne togavirus enzyme immunohistochemistry.[26] Only a few reports exist which detailed the use of this techniques before 1980.[8,38] The virus may be identified directly by monoclonal antibody-enzyme conjugates within a few hours after injection without the need of extensive neutralization tests, and prior to the appearance of CPE (2 to 3 days).[15] This approach also offers the possibility to use invertebrate cells for the direct screening of arbovirus infections. Moreover, slides can be stored for later use. Localization and sequestration of viral antigens is possible at both light and electron microscope levels,[25] and the usefulness of this approach had been extended to the quantification of the antiviral effect of interferon.[44] The immunoenzymatic techniques proved in all cases to be more convenient and more objective than the plaque reduction assays.

In the original fluorescent-antibody test (used to detect an antigen to Colorado tick fever), BHK-21 cells were inoculated with virus and removed 24 hr later from the culture flasks after trypsinization and diluted in PBS to yield 2×10^7 cells per mℓ. Smears of 50 mℓ of cells are placed on microscope slides (e.g., onto multispot slides; Wellcome, Flow) air dried, fixed for 10 min in acetone at room temperature, air dried, and stored at $-70°C$.[15,36] It was discovered that periodate-lysine-paraformaldehyde fixation[30] usually used for electron microscopy yielded a less intensive but more specific staining.[11] A marked difference between antigenic group A (alphaviruses) and antigenic group B (flaviviruses) viruses is the staining observed with the cellular membrane since only the alphaviruses bud from the cell membrane.

Nonspecific staining may be due to methodological nonspecificity or to immunological nonspecificity. The latter source of background staining is important for alphaviruses and flaviviruses due to the close serological relationships among them, whereas methodological nonspecificity is mostly due to endogenous enzyme activity, nonimmunological adsorption of antiserum or methodological errors (insufficient washing, incorrect buffer, pH, substrate concentration). Immunological nonspecificity is generally localized in a specific pattern and is often difficult to eliminate, whereas methodological nonspecificity tends to give a diffuse staining evenly distributed over the preparation. Immunological nonspecificity for alphaviruses or flaviviruses is mostly due to the shared reactivity of the C antigen to different antisera, though cross-reactivity among different antigens (C or envelope) may occur. These considerations are very important when using monoclonal antibodies probably more so than for any other family of viruses.

VI. IMMUNOENZYMATIC SCREENING TECHNIQUES FOR ARBOVIRUSES

The rapid diagnosis of arboviruses is of eminent interest for physicians. Encephalitides may have a very similar clinical presentation, making rapid laboratory diagnosis particularly important.[2]

Some of the enzyme immunoassays (EIA) are very similar to the methods used in enzyme immunohistochemistry, the sole difference being a soluble end product. For example, van Tiel et al.[44] have grown L-cells directly in microtitration plates, infected the cells with Semliki Forest virus, and fixed the cells with 50 $\mu\ell$ of 0.05% glutaraldehyde per well. The fixed monolayers were then washed with PBS, followed by the immunodetection of the antigen. The use of a substrate yielding a soluble product allows quantification with a photometer.

In contrast to hemagglutination-inhibition tests, EIA always detect the highest antibody activity to homologous viruses after primary infection of guinea pigs with flaviviruses.[21] In sequential infections with two flaviviruses they noted that antibody activity after the secondary infection was always higher to the first infecting virus than to the second infecting virus. This stresses the usefulness of EIA for the detection of the "original antigenic sin" phenomenon. Inouye et al.[21] postulated that hemagglutination-inhibition titers to a *Flavivirus* does not reflect the concentration and avidity of antibodies to that virus (hemagglutination-inhibition titers may be higher to heterologous virus).

Novel capture IgM EIA systems have been developed with distinct advantages over conventional EIA.[22,35] Few publications appeared on the use of specific IgM antibody detection for the diagnosis of acute *Flavivirus* infections prior to these developments. Earlier it was necessary to separate IgM from IgG by tedious sucrose gradient methods or by decreasing the avidity by mercaptoethanol treatment.[13,14,19,20,34] With the capture methods, however, IgM can be selectively bound to the solid phase coated with anti-IgM antibodies. After the binding of IgM, enzyme-labeled antigen (or antigen and labeled antibodies) will reveal antibodies among these IgM molecules. Recently this approach has been applied successfully for several flaviviruses such as Japanese encephalitis, tick-borne encephalitis, Dengue, and West Nile virus.[7,19,35] Not only does IgM pinpoint an acute infection, but it also seems to increase the specificity of the test.

VII. CONCLUSION

In vitro immunoenzymatic detection and screening of *Alpha-* or *Flavivirus* antigens of antibodies will be increasingly applied despite earlier problems with shared- and cross-reactivity. The significant advances in EIA and the rapidly growing availability of specific monoclonal antibodies makes it possible to identify a certain agent without the ambiguity of the hemagglutination-inhibition or complement-fixation methods or the expense and slowness of the virus neutralization methods. Hitherto only a few reports have appeared on EIA of arboviruses, though about 1000 papers are published annually with the mention of EIA in their title. However, recent advantages make EIA superior to the classical methods and a widespread use can be anticipated.

ACKNOWLEDGMENTS

This research work was supported by grant A-3746 to Professor E. Kurstak from Natural Sciences and Engineering Council, Canada.

REFERENCES

1. **Atkins, G. J., Sheahan, B. J., and Dimmock, N. J.,** Semliki Forest virus infection of mice: a model for genetic and molecular analysis of viral pathogenecity, *J. Gen. Virol.,* 66, 395, 1985.
2. **Balfour, H. H., Siem, R. A., Bauer, H., and Quie, P. G.,** California arbovirus (La Crosse) infections. I. Clinical and laboratory findings in 66 children, *Pediatrics,* 52, 680, 1973.
3. **Boere, W. A. M., Benaissa-Trouw, B. J., Harmsen, M., Kraaijeveld, C. A., and Snippe, H.,** Neutralizing and non-neutralizing monoclonal antibodies to the E_2 glycoprotein of Semliki Forest virus can protect mice from lethal encephalitis, *J. Gen. Virol.,* 64, 1405, 1983.
4. **Boere, W. A. M., Harmsen, T., Vinje, B., Benaissa-Trouw, B. J., Kraaijeveld, C. A., and Snippe, H.,** Identification of distinct antigenic determinants on Semliki Forest virus by using monoclonal antibodies with different antiviral activities, *J. Virol.,* 52, 575, 1984.
5. **Brandt, W. E., Buescher, E. L., and Hetrick, F. M.,** Production and characterization of arbovirus antibody in mouse ascetic fluid, *Am. J. Trop. Med. Hyg.,* 16, 339, 1967.

6. Brinton, M. A., Analysis of intracellular West Nile virus particles produced by cell cultures, *J. Virol.*, 46, 860, 1983.
7. Burke, D. S. and Nisalak, A., Detection of Japanese encephalitis virus immunoglobulin M antibodies in serum by antibody capture radioimmunoassay, *J. Clin. Microbiol.*, 15, 353, 1982.
8. Cantazaro, P. J., Brandt, W. E., Hogrefe, W. R., and Russell, P. K., Detection of Dengue cell-surface antigens by peroxidase-labeled antibodies and immune cytolysis, *Infect. Immun.*, 10, 381, 1974.
9. Casals, J., Viruses: the versatile parasites. I. The arthropod-borne group of animal viruses, *Trans. N.Y. Acad. Sci.*, (Series 2), 19, 219, 1956.
10. Casals, J., Togaviridae: Alphavirus, in *CRC Handbook Series in Clinical Laboratory Science*, Vol. 1, Seligson, D., Ed., CRC Press, Boca Raton, Fla., 1978, 183.
11. Charpentier, J., Garzon, S., and Kurstak, E., Detection des antigenes de l'enveloppe du virus Chikungunya par la technique d'immunoperoxydase, *Ann. Virol. (Inst. Pasteur, Paris)*, 133E, 233, 1982.
12. Cunningham, S. and Nutter, J. E., Eds., National Institute of Allergy and Infectious Disease Catalog of Research Reagents 1975—77, Publ. No. (NIH) 75-899, Department of Health, Education and Welfare, Bethesda, Md., 1975.
13. Dittmar, D., Cleary, T. J., and Castro, A., Immunoglobulin G and M-specific enzyme-linked immunosorbent assay for detection of Dengue antibodies, *J. Clin. Microbiol*, 9, 498, 1979.
14. Edelman, R. and Pariyanonda, A., Human immunoglobulin M antibody in the serodiagnosis of Japanese encephalitis virus infections, *Am. J. Epidemiol.*, 98, 29, 1973.
15. El Mekki, A. A. and van der Groen, G., A comparison of indirect immunofluorescence on electron microscopy for the diagnosis of some haemorrhagic viruses in cell culture, *J. Virol. Methods*, 3, 61, 1981.
16. Emmons, R. W., Dodero, D. V., Devlin, V., and Lennette, E. H., Serologic diagnosis of Colorado tick fever. A comparison of complement-fixation, immunofluorescence and plague-reduction methods, *Am. J. Trop. Med. Hyg.*, 18, 796, 1969.
17. Fenner, F., The classification and nomenclature of viruses. Summary of results of meetings of the international committee on taxonomy of virus in Madrid, September 1975, *J. Gen. Virol.*, 31, 463, 1976.
18. Garoff, H., Kondor-Koch, C., and Riedel, H., Structure and assembly of alphaviruses, *Curr. Top. Microbiol. Immunol.*, 99, 1, 1982.
19. Heinz, F. X., Roggendorf, M., Hofmann, H., Kunz, C., and Deinhardt, F., Comparison of two different enzyme immunoassays for detection of immunoglobulin M antibodies against tick-borne encephalitis virus in serum and cerebrospinal fluid, *J. Clin. Microbiol.*, 14, 141, 1981.
20. Hofmann, H., Frisch-Niggemeyer, W., and Heinz, F., Rapid diagnosis of tick-borne encephalitis by means of enzyme linked immunosorbent assay, *J. Gen. Virol.*, 42, 505, 1979.
21. Inouye, S., Matsuno, S., and Tsurukubo, Y., "Original antigenic pin" phenomenon in experimental flavivirus infections of guinea pigs: studies by enzyme-linked immunosorbent assay, *Microbiol. Immunol.*, 28, 569, 1984.
22. Jamnback, T. L., Beaty, B. J., Hildreth, S. W., Brown, K. L., and Gundersen, C. B., Capture immunoglobulin M system for rapid diagnosis of La Crosse (California encephalitis) virus infections, *J. Clin. Microbiol.*, 16, 577, 1982.
23. Kennedy, S. I. T., in *The Togaviruses*, Schlesinger, R. W., Ed., Academic Press, New York, 1980, 343.
24. Kurstak, E., Tijssen, P., Kurstak, C., and Morisset, R., Progress in the application of new immunoenzymatic methods in virology, *Ann. N.Y. Acad. Sci.*, 254, 369, 1975.
25. Kurstak, E., Tijssen, P., and Kurstak, C., Immunoperoxidase technique in diagnostic virology and research: principles and application, in *Comparative Diagnosis of Viral Diseases*, Vol. 2, Kurstak, E. and Kurstak, C., Eds., Academic Press, New York, 1977, 403.
26. Kurstak, E., Tijssen, P., and van den Hurk, J., Kurstak, C., and Morisset, R., Detection by immunoperoxidase and ELISA of Arbovirus antigens and antibodies in *in vivo* and *in vitro* systems, in *Invertebrate Systems in vitro*, Kurstak, E., Maramorosch, K., and Dubendorfer, A., Eds., Elsevier/North Holland, Amsterdam, 1980, 365.
27. Kurstak, E., Tijssen, P., and Kurstak, C., Recent advances in immunological and biochemical diagnostic virology, in *Control of Virus Diseases*, Kurstak, E. and Marusyk, R. G., Eds., Marcel Dekker, New York, 1984a, 477.
28. Kurstak, E., Tijssen, P., and Kurstak, C., Enzyme immunoassays applied in virology: reagents preparation and interpretation, in *Applied Virology*, Kurstak, E., Ed., Academic Press, New York, 1984b, 479.
29. Matthews, R. E. F., Classification and Nomenclature of Viruses, 4th Report, Karger, Basel, 1982.
30. McLean, I. W. and Nakane, P. K., Periodate-lysine-paraformaldehyde fixative. A new fixative for immunoelectron microscopy, *J. Histochem. Cytochem.*, 22, 1077, 1974.

31. Monckton, R. P. and Westaway, E. G., Restricted translation of the genome of the flavivirus Kunjin in vitro, *J. Gen. Virol.*, 63, 227, 1982.
32. Palmer, D. F., Kaufman, L., Kaplan, W., and Cavallero, J. J., *Serodiagnosis of Mycotic Diseases,* Charles C Thomas, Springfield, Ill., 1977.
33. Rosen, L. and Gubler, D., The use of mosquitoes to detect and propagate Dengue virus, *Am. J. Trop. Med. Hyg.,* 23, 1153, 1974.
34. Schmitz, H., Improved detection of virus-specifc IgM antibodies. Elimination of non-specific IgM binding, *J. Gen. Virol.,* 40, 459, 1978.
35. Schmitz, H. and Emmerich, P., Detection of specific immunoglobulin M antibody to different *Flavivirus* by use of enzyme-labeled antigens, *J. Clin. Microbiol.,* 19, 664, 1984.
36. Shope, R. E., in *Manual of Clinical Microbiology,* Lennette, E. H., Ballows, A., Hausler, W. J., and Shademy, H. J., Eds., American Society of Microbiology, Washington, D. C., 1985, 785.
37. Simuzu, B., Inhibition of host cell macromolecular synthesis following Togavirus infection, *Compr. Virol.,* 19, 465, 1984.
38. Stollar, V., Harrak, K., Thomas, V., and Sarver, N., in *Arctic and Tropical Viruses,* Kurstak, E., Ed., Academic Press, New York, 1979, 277.
39. Stollar, V., Togaviruses in outlined arthropod cells. Replication strategies of the single stranded RNA-viruses, in *The Togaviruses,* Schlesinger, R. W., Ed., 1980, 583.
40. Strauss, E. G., Rice, C. M., and Strauss, J. H., Sequence coding for the alphavirus nonstructural proteins is interrupted by an opal termination codon, *Proc. Natl. Acad. Sci. U.S.A.,* 80, 5271, 1983.
41. Tijssen, P. and Kurstak, E., Highly efficient and simple methods for the preparation of peroxidase and active peroxidase antibody conjugates for enzyme immunoassays, *Anal. Biochem.,* 136, 451, 1984.
42. Trent, D. W. and Qureshi, A. A., Structural and nonstructural proteins of Saint-Louis encephalitis virus, *J. Virol.,* 7, 379, 1971.
43. van Tiel, F. H., Boere, W. A. M., Vinjé, J., Harmsen, T., Benaissa-Trouw, B. J., Kraaijeveld, C. A., and Snippe, H., Detection of Semliki Forest virus in cell culture by use of an enzyme immunoassay with peroxidase-labeled monoclonal antibodies specific for glycoproteins E_1 and E_2, *J. Clin. Microbiol.,* 20, 387, 1984.
44. van Tiel, F. H., Boere, W. A. M., Harmsen, T., Benaissa-Trouw, B. J., Kraaijeveld, C. A., and Snippe, H., Enzyme immunoassay of interferon with peroxidase-labelled virus-specific monoclonal antibodies, *J. Gen. Virol.,* 66, 1353, 1985.
45. Webb. P. A., Johnson, K. M., Mackenzie, R. B., and Kuns, M. L., Some characteristics of Machupo virus, causative agent of bolivian hemorrhagic fever, *Am. J. Trop. Med. Hyg,* 16, 531, 1967.
46. Weiss, B. and Schlesinger, S., Defective interferon particles do not interfere with the homologous virus obtained from persistently infected BHK cells but do interfere with Semliki Forest virus, *J. Virol.,* 37, 840, 1981.
47. Wengler, G. and Wengler, G., Terminal sequences of the genome and replicative-form RNA of the flavivirus West Nile virus: absence of poly(A) and possible role in RNA replication, *Virology,* 113, 549, 1981.
48. Westaway, E. G., Replication of flaviviruses, in *The Togaviruses,* Schlesinger, R. W., Ed., Academic Press, New York, 1980, 531.
49. Kurstak, E., Progress in enzyme immunoassays: production of reagents, experimental design, and interpretation, *Bull. W.H.O.,* 63(4), 793, 1985.
50. Kurstak, E., *Enzyme Immunodiagnosis,* Academic Press, New York, 1986, 235 pp.

Arthropod Cell Cultures as Systems for the Isolation and Assay of Arboviruses

Chapter 6

APPLICATION OF *AEDES PSEUDOSCUTELLARIS* (AP-61) CELLS TO ARBOVIRUS ISOLATION AND IDENTIFICATION

C. J. Leake and M. G. R. Varma

TABLE OF CONTENTS

I. INTRODUCTION

Virus isolation is the starting point for all arbovirus studies, and this may be amply illustrated by considering the Japanese encephalitis virus as an example. The incidence of this virus has resurged over much of its known range over the past decade and it has also appeared in new geographic areas such as North India and Nepal. Possible explanations for this altered epidemiological pattern may be that new strains of the virus have evolved, strains against which currently available vaccines may be less effective than desired, or that new vector species may have become important. Answers to these questions and many others can only be obtained by studying an assembly of representative geographic and temporal strains of the virus isolated from different sources. To achieve this, systems are required that are not only sensitive to a wide range of viruses, but can also be used both in the laboratory and under difficult field conditions to isolate viruses from mosquito pools, human or animal blood, body fluids, or infected tissues.

In this chapter we describe the application of the *Aedes pseudoscutellaris* (LSTM-AP-61) cell line[1] for the isolation and identification of the three most important arboviruses in the world, i.e., Dengue (DEN), Japanese encephalitis (JE), and yellow fever (YF). These studies were performed in a number of different laboratories and by a variety of techniques.

II. VIRUS ISOLATION AND IDENTIFICATION

A. Dengue Viruses

We first described syncytial cytopathic effect with Dengue-2 virus in *Aedes pseudoscutellaris* cells (Figure 1 A,B) in 1974,[1] and this system was then used by Race et al.[2,3] for isolation studies based at the Caribbean Epidemiology Centre (CAREC) in Trinidad. Flasks of cells were prepared at CAREC, flown to Dominica, inoculated with serum samples, and returned to Trinidad. 238 DEN virus strains were isolated from 664 human sera tested. An analysis of case data in relation to serological results showed that as the hemagglutinating antibody (HAI) titer of the sera rose, the isolation success fell with 33% success (60 out of 182) with HAI titers of <1:40 to 11% (5 out of 44 with HAI titers >1:40). In relation to day of illness isolation, success was 59% (60 out of 101) on day 1 falling to only 21% (22 out of 103) 5 days or longer into the illness.

DEN types 1 and 3 were identified by complement-fixation tests on tissue culture fluids frequently giving monotypic reactions, although disrupted cell suspensions gave more cross-reactions between the four serotypes. Using this technique, a presumptive diagnosis could often be made 4 to 6 days after inoculation. In addition to virus isolates from human sera, one isolate was made by dropping blood from a fingerprick at a health center directly into a flask of AP-61 cells; isolates were also made from mosquito pools.

Subsequently, Hebert et al.[4] used AP-61 cells grown in 4-chamber glass LAB-TEK slides to isolate 58 strains of DEN. Instead of relying on cytopathic effect, they employed a one-step detection and identification procedure by fixing the slides in cold acetone and staining the cells in individual chambers by an indirect fluorescent antibody technique using mouse immune sera raised against each of the 4 DEN serotypes. A total of 29 isolates were made from 1:20 dilutions of human serum, 2 from monkey sera, and 3 from mosquito pools including all 4 serotypes from human sera, and 3 of the 4 serotypes from the mosquito pools.

In a comparative study in Bangkok in 1983[5] we compared the three principal mosquito cell lines, AP-61 cells, C6/36 *Aedes albopictus* cells,[6] and TRA-284-SF *Toxorhynchites amboinensis* cells[7] for virus isolation from plasma samples from Dengue

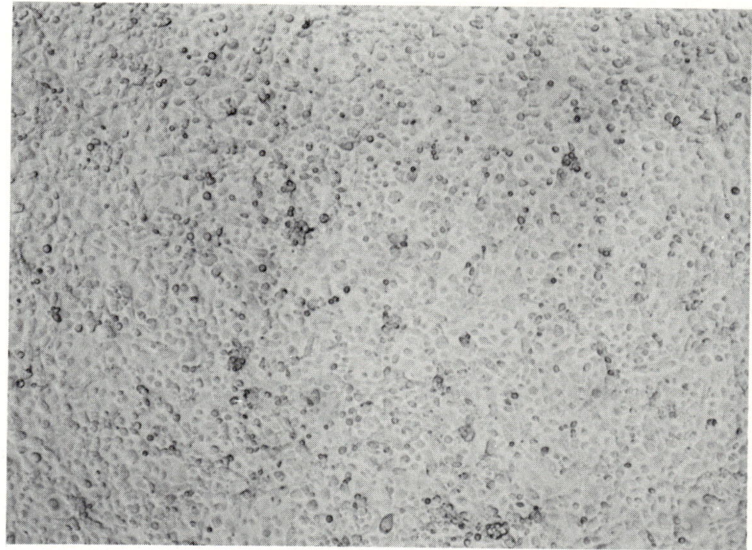

A

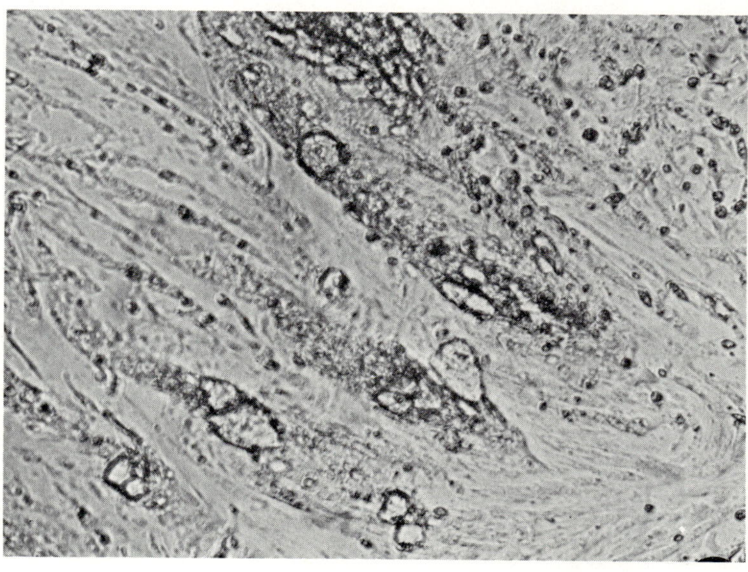

B

FIGURE 1.   A. Control AP-61 cells; B. syncytium formation with Dengue-2 virus; and C. focal cytopathic effect with yellow fever virus.

hemorrhagic fever patients. A total of 77 samples were inoculated onto cells grown in plastic flasks and after 10 days of incubation the cells were resuspended and spotted onto teflon-coated spot slides, air dried, and fixed in cold acetone. Rapid screening was then carried out by indirect immunofluorescence using a *Flavivirus* group-specific monoclonal antibody or DEN type-specific monoclonal antibodies.[8] This method gave clearly positive results, but resolution was relatively poor (Figure 2A) and it was found

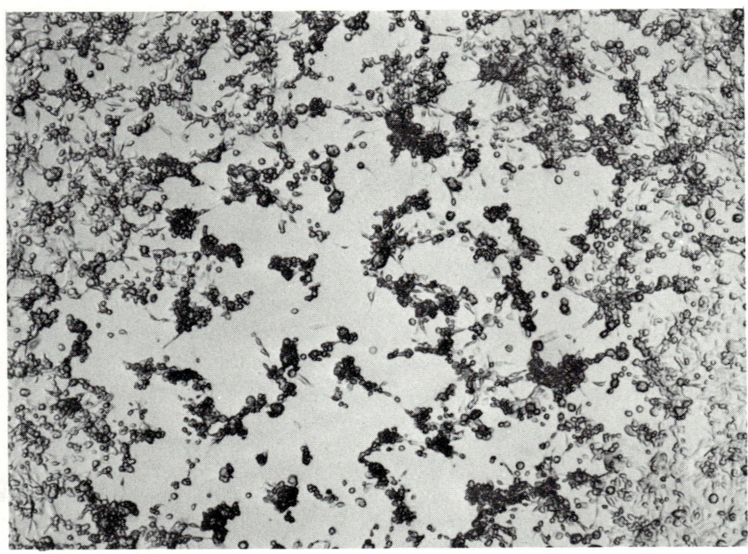

FIGURE 1C

that much better resolution could be obtained (Figure 2B) by growing virus-infected cells on glass microscope coverslips in petri dishes for 2 to 5 days prior to fixation and staining.

Some of our results summarized in Table 1 show that the TRA-284-SF cells isolated 21 viruses from the 77 samples compared to 16 on AP-61 cells and only 8 on C6/36 cells. However, the TRA-284-SF cells were also the slowest growing and most difficult to handle of the 3 cell lines which may restrict their widespread use. In addition to showing quite a high isolation rate and being easy to handle, the AP-61 cells showed the lowest level of plasma toxicity with 1:4 dilutions of plasma inoculated. Only 3 of 77 samples were recorded as toxic on AP-61 cells compared to 25 out of 77 for C6/36 cells. As in the Caribbean study, the HAI titer of the sample had a marked effect on isolation success (Table 1) with no cell culture isolates being made from plasma samples with HAI titers of 2560 or greater.

B. Japanese Encephalitis Virus

Although we originally demonstrated syncytial CPE in AP-61 cells with strains of JE virus in 1974[1] and successfully isolated 14 strains of JE from mosquito pools collected in Sarawak,[9] it was not until 1983 that AP-61 cells were used in Southeast Asia for field isolation of JE.

In studies on the epidemiology of JE in northern Thailand, 23 strains of JE were reisolated on AP-61 cells from mosquito pools collected in light traps. Viral antigen was detected by a rapid JE-specific antigen capture immunoassay on infected culture fluids.[10] Further, a collection of 56 flaviviruses isolated on C6/36 *Aedes albopictus* cells were typed on first passage on AP-61 cells using *Flavivirus* and JE-specific monoclonal antibodies in indirect immunofluorescence tests. The *Flavivirus* group-specific monoclonal antibody correctly identified all the flaviviruses, whereas some unexpected cross-reactions with the Tembusu virus were observed with 1 of the JE specific monoclonal antibodies.[10]

Additional studies in both northern and southern Thailand have isolated JE virus strains from sentinel pig sera using AP-61 cells.[11,12] Parallel inoculation of C6/36 cells

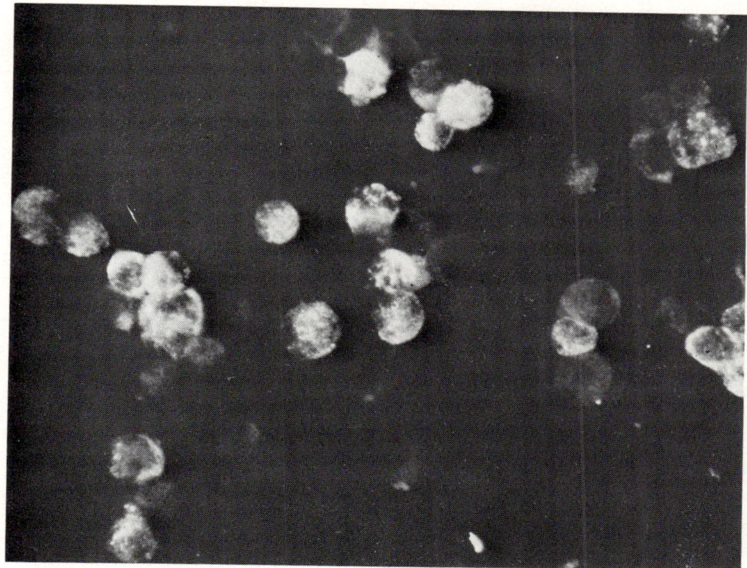

A

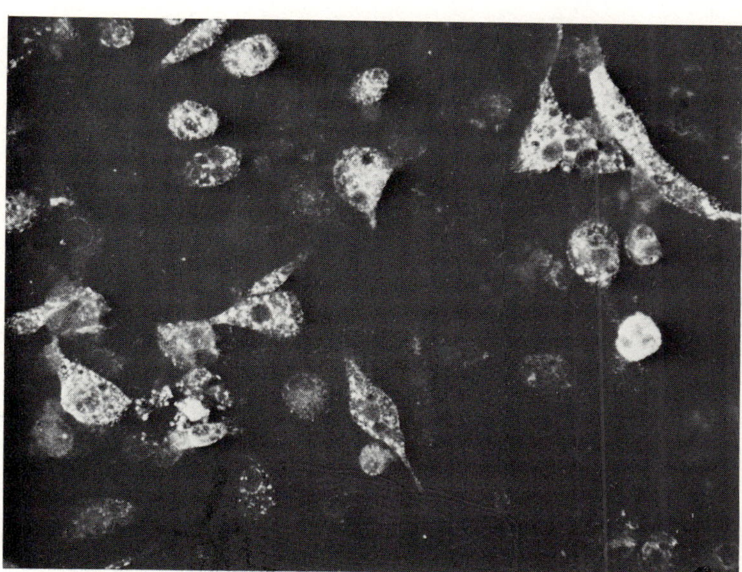

B

FIGURE 2.   A. AP-61 cells infected with Dengue-2 virus: spot slide preparation. Indirect immunofluorescence with *Flavivirus* group-reactive monoclonal antibody. Note rounded cell outlines and lack of intracellular detail; B. AP-61 cells infected with Japanese encephalitis virus: cover slip preparation. Indirect immunofluorescence with *Flavivirus* group-reactive monoclonal antibody. Cells are flattened revealing excellent intracellular detail.

Table 1

ISOLATION OF DENGUE (DEN) VIRUSES ON MOSQUITO CELL LINES FROM SELECTED DENGUE HEMORRHAGIC FEVER PATIENTS

	Strain isolated	Mosquito inoculation	C6/36	Cell lines LSTM-AP-61	TRA-284-SF
Group 1	DEN 1	3	1	2	3
HAI[a] <10, 10, or 20	DEN 2	5	3	5	4
	DEN 3	2	0	2	1
	DEN 4	0	0	0	0
Group 2	DEN 1	1	0	0	1
HAI 40, 80, or 160	DEN 2	3	2	1	3
	DEN 3	1	0	1	1
	DEN 4	1	0	1	1
Group 3	DEN 1	0	0	0	0
HAI 320, 640, or 1280	DEN 2	7	1	3	4
	DEN 3	1	0	0	1
	DEN 4	4	1	1	2
Group 4	DEN 1	3	0	0	0
HAI 2560, 5120, or 10,240+	DEN 2	7	0	0	0
	DEN 3	0	0	0	0
	DEN 4	1	0	0	0
Total		39	8	16	21

[a]   HAI: hemagglutinating antibody titer.

frequently showed toxic effects of pig sera, whereas the AP-61 cells were relatively unaffected isolating many more JE virus strains.

Perhaps the most important facet of this epidemiological work was the isolation of JE virus strains from clinical specimens collected during the 1983 epidemic in northern Thailand.[13] AP-61 cells were prepared in Bangkok and shipped by road 350km to the provincial hospital at Kamphaengphet where they were inoculated with brain autopsy or brain biopsy material from fatal patients or cerebrospinal fluid from all patients. After inoculation by a laboratory technician, the cells were returned to Bangkok by road having been maintained throughout this period under variable ambient temperatures. Once back in the laboratory the cells were examined by monoclonal antibody immunofluorescence tests 10 to 14 days after inoculation. JE virus strains were isolated from most areas of the brains of seven fatal patients. Based on the frequency of virus isolation, the most heavily affected areas were the thalamus (8/14), followed by the cerebellum (6/14), medulla (6/14), frontal (6/14), and occipital lobes (6/14) with virus isolates from the pons being less frequent, (3/14). A single strain was isolated from 4 brain biopsies and in addition 5 virus isolates were obtained from cerebrospinal fluid specimens, all from ultimately fatal cases. The presence of virus in the cerebrospinal fluid therefore appeared to be an ominous prognostic sign to eventual disease outcome.

C. Yellow Fever Virus

In comparison with DEN and JE viruses, studies with YF virus have been more limited. After describing the first mosquito cell isolation of YF virus from mosquito pools and infected human liver suspension and sentinel rhesus monkey serum from Panama,[14] in Trinidad in 1979 we isolated this virus using cytopathic effect as a criterion (Figure 1 A,C) and performed plaque neutralizations in AP-61 cells under carboxymethylcellulose overlays.[15] The cytopathic effect on AP-61 cells was very different from that seen with DEN or JE viruses, being cytocidal in nature. Cell death was only seen with wild strains of the virus and not with the 17D vaccine strain.

Recently AP-61 cells have been used at the Pasteur Institute in Dakar to isolate YF strains from the Burkina Faso (formerly Upper Volta) epidemic in 1983. The AP-61 cells (mosquito isolates 26/120, blood isolates 5/17, and liver isolates 11/13) proved slightly superior to intrathoracic inoculation of mosquitoes (mosquito isolates 24/120, blood isolates 5/17, and liver isolates 7/13) and far superior to an antigen capture enzyme linked immunoassay (mosquito isolates 2/120, blood isolates 3/17, and liver isolates 8/13).[16] In addition, 25 strains were isolated from mosquito pools on AP-61 cells during an epizootic in Eastern Senegal late in 1983 and in an interesting extension of these studies at the Pasteur Institute on the Ivory Coast, YF-IgM immunocomplexes previously negative for virus isolation yielded 36 isolates from 99 samples on C6/36 *Aedes albopictus* cells after the complexes were dissociated with dithiothreitol, although no comparisons were made with the AP-61 system.[16]

III. CONCLUSIONS

Isolation of arboviruses in *Aedes pseudoscutellaris* cells has been a process of steady evolution from initial laboratory demonstration to field application in areas as far apart as the Caribbean, Africa, and Southeast Asia. The cells now have a proven record for being robust, having been shipped by air in the Caribbean and shipped 700 km by road in Thailand unattended and under ambient conditions, then inoculated by technical staff rather than specialist scientists. Virus isolations have been made from mosquito pools, monkey and pig sera, whole human blood, serum and plasma from human brain autopsies, human brain biopsies, cerebrospinal fluid, and liver suspensions. The resilience of the AP-61 cells is apparent, as many of these samples have been shown to be toxic to other systems.

Furthermore, the flexibility of the AP-61 cell system is emphasized by the fact that the cells will grow readily in commercially available tissue culture media as well as in specialized formulations. They have been grown on glass bottles, and on a variety of tissue culture plastics and will adhere to glass spot slides, microscope coverslips, and LAB-TEK slides for special treatments.

Similarly, virus identification methods have been varied including cytopathology, plaque neutralization, hyperimmune and monoclonal fluorescent antibody staining on cells, and complement-fixation tests and specific monoclonal antibody enzyme linked immunoassay tests on infected tissue culture fluids. It is now imminent that cDNA probes will be applied to this system.

Finally, the examples quoted here represent only a few of the viruses that will grow in these cells,[17] some of which are discussed elsewhere in this book. Indeed, we have described the isolation of the alphaviruses Sindbis and Getah in AP-61 cells[17] based on syncytial cytopathic effect similar to DEN, YF, and West Nile viruses. The *Bunyavirus* Ingwavuma has also been isolated on AP-61 cells, as well as several other untyped viruses[18] and many others can be expected in the future. If we are to establish country-wide or regional arbovirus surveillance systems, the *A. pseudoscutellaris* cell line would appear to be an excellent overall choice.

ACKNOWLEDGMENTS

The early laboratory work described in this chapter was performed under successive grants from the Medical Research Council of Great Britain. During this period travel funds for C. J. Leake to perform field work on the YF virus in Trinidad were provided by the Wellcome Trust. Field and laboratory studies on JE and DEN viruses were supported by a grant from the Wellcome Trust while C. J. Leake was based at the Department of Virology, U.S. Component of the Armed Forces Research Institute of Medical Sciences in Bangkok.

REFERENCES

1. Varma, M. G. R., Pudney, M., and Leake, C. J., Cell lines from larvae of *Aedes (Stegomyia) malayensis* (Colless) and *Aedes (S) pseudoscutellaris* (Theobald) and their infection with some arboviruses, *Trans. R. Soc. Trop. Med. Hyg.*, 68, 374, 1974.
2. Race, M. W., Fortune, R. A. J., Agostini, C., and Varma, M. G. R., Isolation of dengue viruses in mosquito cells under field conditions, *Lancet*, 1, 48, 1978.
3. Race, M. W., Williams, M. C., and Agostini, C. F. M., Dengue in the Caribbean: virus isolations in a mosquito (*Aedes pseudoscutellaris*) cell line, *Trans. R. Soc. Trop. Med. Hyg.*, 73, 18, 1978b.
4. Hebert, S. J., Bowman, K. A., Rudnick, A., and Burton, J. J. S., A rapid method for the isolation and identification of Dengue viruses employing a single system, *Malays. J. Pathol.*, 3, 67, 1980.
5. Leake, C. J., Nisalak, A., and Burke, D. S., Comparative isolation of Dengue viruses by mosquito inoculation and on three mosquito cell lines, Proc. Int. Conf. on Dengue/Dengue Haemorrhagic Fever, Kuala Lumpur, Malaysia, 1—3 September 1983, 437, 1983.
6. Igarashi, A., Isolation of a Singh's *Aedes albopictus* cell clone sensitive to dengue and chikungunya viruses, *J. Gen. Virol.*, 40, 531, 1978.
7. Kuno, G., Dengue virus replication in a polyploid mosquito cell culture grown in serum-free medium, *J. Clin. Microbiol.*, 16, 851, 1982.
8. Henchal, E., McCown, J. M., Seguin, M. C., Gentry, M. K., and Brandt, W. E., Rapid identification of Dengue virus isolates by using monoclonal antibodies in an indirect immunofluorescence assay, *Am. J. Trop. Med. Hyg.*, 32, 164, 1983.
9. Pudney, M., Leake, C. J., and Varma, M. G. R., Replication of arboviruses in arthropod *in vitro* systems, in *Arctic and Tropical Arboviruses*, Kurstak, E., Ed., Academic Press, New York, 1979.
10. Leake, C. J., Ussery, M. A., Nisalak, A., Hoke, C. H., Andre, R. G., and Burke, D. S., Unpublished data, 1984.
11. Burke, D. S., Tingpalapong, M., Ward, G. S., and Leake, C. J., Unpublished data, 1983.
12. Burke, D. S., Ussery, M. A., Elwell, M. R., Nisalak, A., Leake, C. J., and Laorakpongse, T., Isolation of Japanese encephalitis virus strains from sentinel pigs in Northern Thailand, *Trans. R. Soc. Trop. Med. Hyg.*, In press, 1985.
13. Leake, C. J., Burke, D. S., Nisalak, A., and Hoke, C. H., Unpublished data, 1983.
14. Varma, M. G. R., Pudney, M., Leake, C. J., and Peralta, P. H., Isolations in a mosquito (*Aedes pseudoscutellaris*) cell line (Mos 61) of yellow fever virus strains from original field material, *Intervirology*, 6, 50, 1976.
15. Leake, C. J., Unpublished data, 1979.
16. *Wld. Hlth. Org. Wkly, Epidem, Rec.*, Yellow fever in 1983, 59, 329, 1984.
17. Pudney, M., Leake, C. J., and Buckley, S. M., Replication of arboviruses in arthropod *in vitro* systems: an overview, in *Invertebrate Cell Culture Applications*, Maramorosch, K. and Mitsuhashi, J., Eds., Academic Press, 1982, 159.
18. Leake, C. J., Unpublished data, 1984.

Chapter 7

TICK CELL LINES FOR THE ISOLATION AND ASSAY OF ARBOVIRUSES

M. Pudney

TABLE OF CONTENTS

I. INTRODUCTION

Despite the fact that ticks are such diverse and versatile transmitters of infectious agents to man and animals and that continuous cell lines have obvious benefits for studies of vector-borne disease agents, vector-pathogen interrelationships, and control, few investigators have succeeded in establishing cell lines from them.[1] Ticks are second only in importance to mosquitoes as vectors of arboviruses and it was the requirements of virologists for invertebrate host cells in vitro that provided the initial impetus for tick cell culture. Until recently the progress of tick cell lines has continued to be associated with arbovirus studies, with relatively little application in the study of other pathogens.

The comparatively late start of acarine cell culture (the first report appeared in 1952,[2] 35 years after insect culture was first attempted), and the early difficulties encountered has meant that this area of endeavor has continued to lag behind the advances made using mosquito cells. It was not until 1975 that we reported the establishment of the first three continuous tick cell lines from *Rhipicephalus appendiculatus*.[3] There has since been a steady increase in the number of tick cell lines, particularly from embryonic cells[1,4,5] and the methods used to initiate cultures have become fairly standardized.[6,7] However, all the cell lines are from ixodid ticks (no cell line from argasid ticks having yet been produced) and limited to the following species: *Rhipicephalus appendiculatus, R. sanguineus, Dermacentor parumapertus, D. variabilis, Haemaphysalis spinigera, H. obesa,* and *Boophilus microplus*.[1]

Early studies of arbovirus growth in tick cells has necessarily been limited. Primary cultures of *Hyalomma dromedarii* were shown to support the replication of 15 arboviruses, all togaviruses by Rehacek,[8] and 4 tick-borne viruses, LI*, QRF, ZIR, and LJN by Pudney[9] and CTF virus multiplied in primary cultures of its vector, *Dermacentor andersoni*.[10] None of these viruses produced a cytopathic effect. This early progress and the potential of these cultures for the cultivation of viruses and rickettsia have been thoroughly reviewed and need not be discussed in detail here.[11]

II. ARBOVIRUS REPLICATION IN TICK CELL LINES

Despite the increasing numbers of tick cell lines being produced, very few of them have been used extensively for comparative studies on a wide range of arboviruses. Those which have are from five different species: *Rhipicephalus appendiculatus* (RA-219, RA-243, and RA-256,[3] *Dermacentor parumpertus* (RML-14),[12,13] *D. variabilis* (RML-15, RML-19),[13] *Haemaphysalis spinigera* (ATC-309),[14] and *Boophilus microplus* (BM-270, BM-256).[15,16] These investigations have been carried out in different laboratories throughout the world. Representing most of the major taxonomic and serological groups, 64 different arboviruses (mostly from mouse brain stocks) have now been examined for their ability to grow in tick cells in vitro, although only two (POW and CHIK) have been tested in all four genera. The results reviewed below are from References 1, 3, 12, 16 to 19, 22 to 24, 31, and 35.

A. Togaviridae; *Alphavirus*

Although all of these viruses are mosquito-borne, 8/9 examined grew in one or more of the tick cell lines. CHIK grew in all except that from *Haemaphysalis spinigera*. Alphaviruses have also been shown to replicate in a wide range of dipteran cell lines, but not in the one lepidopteran line tested[1] (Table 1).

The alphaviruses showed variable patterns of growth in the tick cells, with ONN and

---

* Arboviruses and serogroup abbreviations are given in Tables 1 through 5.

Table 1
ARBOVIRUS GROWTH IN TICK CELL LINES[a]

Family Togaviridae	Abbreviation	Sero group	Vector	*Rhipicephalus appendiculatus*	*Boophilus microplus*	*Dermacentor parumapertus*	*Dermacentor variabilis*	*Haemaphysalis spinigera*
Alphavirus								
Chikungunya	CHIK	A	C/M	+	+	+	+	—
Getah	GET	A	CA/M	+				
Middleburg	MID	A	C/M	—				
Ndumu	NDU	A	C/M	+				
O'Nyong-Nyong	ONN	A	CA/M	+			+	
Semliki Forest	SF	A	CA/M	+	+			
Sindbis	SIN	A	CA/M	+	—			+
Western equine encephalitis	WEE	A	CA/M	+				
Whataroa	WHA	A	C/M	+				

Note: C/M = Culicine mosquito; A/M = Anopheline mosquito; CA/M = Culicine and anopheline mosquito; I/T = Ixodid tick; A/T = Argasid tick; IA/T = Ixodid and argasid tick; Cul. = Culicoides; P. = Phlebotomine; + = Replication; — = No replication; ? = Maintenance at low level; and ( ) = Conflicting results.

[a] Compiled from results given in References 1, 3, 12, 16—19, 22—24, 31, and 35.

WHA reaching fairly low titer in RA243 while SIN and SF grew well. There was a rapid rise in extracellular SF virus one day postinfection (p.i.) in RA243 to over 6 dex PFU/ml, followed by a marked fall, whereas WHA produced a peak titer of 4.2 dex PFU/ml on day 4, but over 3 dex PFU/ml was still being produced extracellularly on day 22. This peak titer of 6.8 dex PFU/ml was also obtained with SF in BM-270 one day p.i., but the drop in extracellular titer was not so rapid and 4 dex PFU/ml was still detectable by 12 days.[19] Similarly, CHIK showed significant growth in the *Dermacentor parumapertus* (RML-14) cell line with a peak titer of 7.5 dex PFU/ml which had fallen to 4.5 PFU/ml by 15 days p.i.[12] Small differences in titer may be due to the fact that in the experiments using RA-243 and BM-270, released virus in the supernatant was measured, whereas in those involving the RML-14 and RML-15 lines total virus was measured by freeze/thawing samples of different days. In both cases virus was titrated by plaque assay using BHK-21, PS, Vero, or the *Xenopus laevis* (XTC-2) cell lines (except for MID and GAN where cytopathic response in XTC-2 cells was used).

B. Flaviviridae; *Flavivirus*

All of the tick-borne viruses tested, 7/7, grew readily in one or more of the cell lines, as did 4/5 of the mosquito-borne flaviviruses. However when tested in other arthropod cells, in general tick-borne flaviviruses would not grow.[1] MOD and RB which have not been shown to be arthropod-borne did not grow in the tick cells (or mosquito cells), supporting the view that arthropods are not involved in their transmission. It would be tempting to speculate that the higher specificity of host range of the tick-borne flavi-

Table 2
ARBOVIRUS GROWTH IN TICK CELL LINES[a]

Family Togaviridae	Abbreviation	Sero group	Vector	Rhipicephalus appendiculatus	Boophilus microplus	Dermacentor parumapertus	Dermacentor variabilis	Haemaphysalis spinigera
Flavivirus								
Dengue 2	DEN-2	B	C/M	−	−			−
Japanese encephalitis	JE	B	CA/M				+	
St. Louis encephalitis	SLE	B	CA/M				+	
West Nile	WN	B	CA/M	+	+	+		−
Yellow fever	YF	B	C/M	(−)	(+)		+	
Kyasunur forst disease	KFD	B	IA/T					+
Langat	LGT	B	I/T	+	+	+	+	
Louping ill	LI	B	I/T		+	+		
Omsk Hemorrhagic fever	OMSK	B	I/T				+	
Powassan	POW	B	I/T	+	+		+	+
Russian spring-summer encephalitis	RSSE	B	I/T			+		
Tyuleniy	TYU	B	I/T		+			
Modoc	MOD	B	None	−		−	−	
Rio Bravo	RB	B	None	−		−		

Note: C/M = Culicine mosquito; A/M = Anopheline mosquito; CA/M = Culicine and anopheline mosquito; I/T = Ixodid tick; A/T = Argasid tick; IA/T = Ixodid and argasid tick; Cul. = Culicoides; P. = Phlebotomine; + = Replication; — = No replication; ? = Maintenance at low level; and ( ) = Conflicting results.

[a]   Compiled from results given in References 1, 3, 12, 16—19, 22—24, 31, and 35.

viruses is indicative of a more highly adapted virus which has envolved in close association with its host in direct comparison to the mosquito-borne flaviviruses such as WN which grows in cell lines from 13 different arthropod species[1] and has been isolated in the field from culicine and anopheline mosquitoes and occasionally ticks (Table 2).

Of the mosquito-borne flaviviruses, WN produced a peak titer of 4 dex PFU/mℓ 3 days p.i. in BM-270 and this level was maintained with a slight decrease until 15 days p.i. when the experiment was terminated. In contrast, growth of WN in RA-243 did not reach a peak (3.6 dex PFU/mℓ) until 10 days p.i. Similarly, the 17D strain of YF grew only poorly in the BM-270 line with a peak titer of 4 dex PFU/mℓ being attained on 9 days p.i. Titer was maintained at this level until the end of the experiment. In the RML-14 cell line both a strain of WN isolated from ticks and one isolated from mosquitoes gave similar growth characteristics, producing peak titers on day 10 of about 6 dex PFU/mℓ despite differences in origin, passage history, and plaque size of the two strains.[12]

Tick-borne flaviviruses grew well, with LGT and LI reaching peak titer of 5.5 and 6.6 dex PFU/mℓ, respectively, by day 4, falling by 2 logs during the remainder of the experiment in RA-243 cells. In BM-270, LGT reached 6.4 dex PFU/mℓ by 6 days and LI 6.3 dex PFU/mℓ by 4 days and titers were maintained above 4 dex PFU/mℓ at

Table 3
ARBOVIRUS GROWTH IN TICK CELL LINES[a]

Family Bunyaviridae	Abbreviation	Sero group	Vector	*Rhipicephalus appendiculatus*	*Boophilus microplus*	*Dermacentor parumapertus*	*Dermacentor variabilis*	*Haemaphysalis spinigera*
Anopheles A	ANA	ANA	A/M	—				
Bunyamwera	BUN	BUN	C/M	?	?			
Germiston	GER	BUN	C/M	—				
California encephalitis	CE	CAL	C/M	—	—			
Tahyna	TAH	CAL	CA/M	—				
Bahig	BAH	TETE	I/T	—				
Sandfly fever (Sicilian)	SFS	PHL	P.	—	—			
Hazara	HAZ	CON	I/T			+	+	
Kaisodi	KSO	KSO	I/T	+				+
Silverwater	SIL	KSO	I/T			?	?	
Dugbe	DUG	NSD	C/M,I/T,Cul.	+	+	+		
Ganjam	GAN	NSD	C/M,I/T	+	+			+
Nairobi sheep disease	NSD	NSD	I/T	+				
Grand Arbaud	GA	UUK	A/T	—				
Uukuniemi	UUK	UUK	I/T	+				
Sakhalin	SAK	SAK	I/T	—				
Bhanja	BHA	Ungrouped	I/T			?		+
Lone star	LS	Ungrouped	I/T	—				
Sunday Canyon	SCA	Ungrouped	A/T			?		

Note: C/M = Culicine mosquito; A/M = Anopheline mosquito; CA/M = Culicine and anopheline mosquito; I/T = Ixodid tick; A/T = Argasid tick; IA/T = Ixodid and argasid tick; Cul. = Culicoides; P. = Phlebotomine; + = Replication; — = No replication; ? = Maintenance at low level; and ( ) = Conflicting results.

[a]   Compiled from results given in References 1, 3, 12, 16—19, 22—24, 31, and 35.

completion on 15 days. Similarly, LGT also grew well in RML-14 with a peak of 5.5 dex PFU/m*l* on day 6 and remaining above 4 dex by 15 days. The other tick-borne flaviviruses grown in RML-14, RSSE, OMSK, and POW all grew to relatively high levels after an initial lag phase, except for POW where an increase in the virus was detected from the day 1. Maximum titer of 4.8 to 6.6 dex PFU/m*l* was obtained between 4 to 10 days p.i. with these viruses.

C. Bunyaviridae

Only 19 of this very large group of viruses have been examined for their growth potential in tick cell lines. Only 1 of 7 isolated from mosquitoes, BUN, multiplied in tick cells and then only to a very low level. In tick cells, 9/13 of the tick-borne bunyaviruses grew, with 2 more being maintained at a low level, whereas they would not grow in mosquito cells,[1] apart from GAN which has also been isolated from mosquitoes (Table 3).

In RA 243, DUG reached 4 dex PFU/m*l* by 6 days and titer was maintained at this level for a further 16 days, whereas GAN reached a slightly higher peak of 5.4 dex

Table 4
ARBOVIRUS GROWTH IN TICK CELL LINES[a]

Family Reoviridae	Abbreviation	Sero group	Vector	Rhipicephalus appendiculatus	Boophilus microplus	Dermacentor parumapertus	Dermacentor variabilis	Haemaphysalis spinigera
Orbivirus								
Colorado tick fever	CTF	CTF	IA/T				+	
Chenuda	CNU	KEM	A/T	—				+
Kemerovo	KEM	KEM	I/T				+	
Nugget	NUG	KEM	I/T	+				
Tribec	TRB	KEM	I/T	+			+	
Wad medani	WM	KEM	I/T					+
Orungo	ORU	Ungrouped	CA/M					+
Arenaviridae								
Pichinde		TCR	I/T	—				
Picornaviridae								
Kawino		Ungrouped	C/M	—				
Rhabdoviridae								
Sawgrass	SAW	SAW	I/T	+				+
Connecticut			I/T	+				
Chandipura	CHP	VSV	P.	—	+		—	
Piry	PIRY	VSV	Unknown	—	?			

Note: C/M = Culicine mosquito; A/M = Anopheline mosquito; CA/M = Culicine and anopheline mosquito; I/T = Ixodid tick; A/T = Argasid tick; IA/T = Ixodid and argasid tick; Cul. = Culicoides; P. = Phlebotomine; + = Replication; — = No replication; ? = Maintenance at low level; and ( ) = Conflicting results.

[a]   Compiled from results given in References 1, 3, 12, 16—19, 22—24, 31, and 35.

PFU/mℓ by 3 days which fell gradually to 3.5 dex PFU/mℓ by 22 days p.i. In RML-14 the peak titer for DUG of 5.4 dex PFU/mℓ was obtained on day 6 and that for HAZ of 5.5 dex PFU/mℓ on day 15. Hazara virus which was inoculated at a concentration too low to detect in Vero cells remained undetected in the RML-14 cells up to day 4. It was first detected on day 6 and thereafter its titer increased rapidly up to 15 days when the experiment was terminated. *Rhipicephalus* cells inoculated with NSD virus yielded a titer of 5 to 6 dex PFU/mℓ, but the growth kinetics are not described.[22]

D. Reoviridae; *Orbivirus*

All of the orbiviruses so far tested (7) have grown in tick cells. This includes one mosquito-borne example, ORU, with the remainder being tick-borne. These viruses will also grow in mosquito cells.[1] Although similar peak titers of 5.1 and 5.3 dex PFU/mℓ were obtained for KEM and TRB in RML-14, the first was not reached until 10 days and the second on 4 days. Whereas these viruses exhibited the typical growth curve terminating in a plateau phase, two strains of CTF virus continued to increase in titer up to day 15 when the was experiment was terminated. At this time maximum yields were 6.5 dex PFU/mℓ (Florio strain) and 5.7 dex PFU/mℓ (strain SS-18)[12] (Table 4).

Table 5

ARBOVIRUS GROWTH IN TICK CELL LINES[a]

Family unclassified	Abbreviation	Sero group	Vector	*Rhipicephalus appendiculatus*	*Boophilus microplus*	*Dermacentor parumapertus*	*Dermacentor variabilis*	*Haemaphysalis spinigera*
Hughes	HUG	HUG	A/T	+		+		
Punta Salinas	PS	HUG	A/T	+				
Soldado	SOL	HUG	A/T	+		+		
Zirqa	ZIR	HUG	A/T	+	+			
Sapphire II		HUG	A/T			—		
Qalyub	QYB	QYB	A/T	+				
Quaranfil	QRF	QRF	A/T	+	+	—		
Midway		NYM	A/T			+		
Dhori	DHO	Ungrouped	I/T			+		
Keterah	KTR	Ungrouped	A/T	+	?			
Casade			I/T					+

Note: C/M = Culicine mosquito; A/M = Anopheline mosquito; CA/M = Culicine and anopheline mosquito; I/T = Ixodid tick; A/T = Argasid tick; IA/T = Ixodid and argasid tick; Cul. = Culicoides; P. = Phlebotomine; + = Replication; — = No replication; ? = Maintenance at low level; and ( ) = Conflicting results.

[a] Compiled from results given in References 1, 3, 12, 16—19, 22—24, 31, and 35.

E. Rhabdoviridae

Only 4 rhabdoviruses have been tested, but all were shown to replicate in tick cells irrespective of their known or suspected vector. These included the tick-borne SAW and Connecticut. Only tissue culture-derived Connecticut virus would infect the tick cells; four attempts with mouse-brain derived material proved negative.[23] This virus will also grow in mosquito cells. In BM-270 the phlebotomine-transmitted Chandipura reached a peak of 4.4 dex PFU/mℓ by day 6, which was maintained at this titer until the end of the experiment. There was little if any growth with PIRY virus whose vector is as yet unknown (Table 4).

F. Unclassified Tick-Borne Arboviruses

Of the 11 viruses of this group, only sapphire II failed to grow in one of the cell lines despite the fact most of them had been isolated from argasid rather than ixodid ticks. None of this group would grow in mosquito cells. The growth kinetics, however, were quite variable. QRF (strain AR1095), although isolated from argasid ticks, produced a peak titer of 5.6 dex CPD$_{50}$/mℓ on days 3 to 7 in RA-243, which gradually fell with 4 dex of virus still being produced 26 days p.i. However, the Ar 1113 strain did not grow in RML-14 cells. QRF also produced a peak titer of 6.5 dex PFU/mℓ on day 3 in the BM-270 cell line, remaining above 4 dex PFU/mℓ by day 15. The RA-243 line was sensitive to infection with QRF virus and could be infected with a 10^{-7} dilution of the seed virus (Table 5).

All of the Hughes group viruses appeared to have a lag phase before extracellular virus could be detected in the RA-243 system; this phase varied from 7 days for ZIR, to 6 days for PS, and 3 days for SOL. This was followed by a gradual slow rise with low peak titer (2.4 dex PFU/mℓ for ZIR, on day 26; 4.0 dex PFU/mℓ for SOL on day 22, and 5.5 dex PFU/mℓ for PS on day 14). However, in the BM-270 cell line there was no lag phase for ZIR and the peak titer of 6.2 was produced by day 3, although it dropped more rapidly than QRF to 2 dex PFU/mℓ by day 15. In RML-14 cells, HUG virus produced maximum titers of 5.5 dex PFU/mℓ by 15 days, after a continual rise from day 3, whereas SOL produced 3.7 dex PFU/mℓ on day 8 and remained at this level. Midway (Nyamanini serogroup) and DHO (serologically ungrouped) also grew well in these cells producing peak titers of 2.5 dex PFU/mℓ and 6.1 dex PFU/mℓ, respectively, on days 15 and 4. No Midway virus was detectable until day 6 when there was a gradual rise in titer until the termination of the experiment on day 15.

G. Nonarboviruses

The arenaviruses are not considered to require an arthropod vector for maintenance although historically they have been closely associated with arboviruses and have been isolated from arthropods. Pichinde virus, although it has been isolated from ticks, showed no apparent growth in RA-243 when assayed by immunofluorescence,[24] confirming its nonarbovirus status. (Similarly, Tacaribe virus did not grow in mosquito cells). The growth of LCM virus in tick primary cultures was reported by Rehacek in 1965,[25] but no further studies have been made since then.

The picornavirus Kawino isolated from mosquito cells in culture[26] did not grow in RA-243 although not surprisingly it did grow in mosquito cells, as it is probably a mosquito picornavirus.[1] As far as this author is aware, no other investigations on the replication of insect viruses in tick cells have been carried out although certain insect iridoviruses and reoviruses will grow in mosquito cells.[1]

Munz et al.[22] were unable to show replication in RA-243 of vertebrate viruses representing a wide range of families, i.e., Reoviridae (Reovirus type 3), Picornaviridae (human poliovirus 1), Poxviridae (vaccinia virus, ectromelia virus, fowl pox virus, and Orf virus), Herpesviridae (Aujeszky virus), and Adenoviridae (bovine adenovirus type 3), confirming the earlier negative results obtained with primary cultures.[1]

Tick cell lines permit growth of both mosquito- and tick-borne arboviruses, indicating that type of vector and absence or presence of essential lipids are not eminent factors in determining infectivity of viruses for tick cells as is the case for mosquito cells[1] (Table 6). This means that the viruses transmitted by culicine and anopheline mosquitoes as well as by argasid and ixodid ticks and possibly phlebotomines (the sandfly-transmitted *Bunyavirus* SFS did not grow in these cells but the phlebotomine transmitted *Rhabdovirus* CHP did grow in one of the lines used), will multiply in tick cells. Culicoides-transmitted viruses have yet to be tested in tick cells.

The susceptibility of cultured tick cells although broad does not extend to those viruses nominally classified as arboviruses, but which apparently do not utilize an intermediate vector. Mod and RB viruses both antigenically allied with the flaviviruses and capable of direct transmission among vertebrate hosts failed to multiply in tick cells. Similarly, the tick-borne *Iridovirus* African swine fever failed to grow in *R. appendiculatus* cells, although the argasid tick *Ornithodoros moubata* is a natural vector.

It is obvious from these studies that there is considerable variation in the growth pattern of a particular virus in various cell lines. Some of the variations in detail may be accounted for by differences in techniques used both for virus growth and assessment in different laboratories (e.g., RA-243 is maintained at 28°C, RML-14 at 29°C, BM-270 at 32°C, and RML-15 and RML-19 at both 27 and 37°C). In RML-14, viruses from ixodid ticks multiplied well or were maintained at high levels, whereas only HUG virus from argasid ticks grew well and others grew poorly or not at all. However, the QRF virus also from argasid ticks grew well in *R. appendiculatus* lines and in the BM-

Table 6
COMPARISON OF ARBOVIRUS GROWTH IN TICK AND MOSQUITO CELL LINES

Virus (genus)	Vector[a]	Tick cell lines[b]	Mosquito cell lines[c]
Alphavirus	Mosquito	8/9	15/15
Flavivirus	Mosquito	4/5	13/15
	Tick	7/7	1/7
	None	0/2	0/3
Bunyavirus	Mosquito	1/7	18/20
	Culcoides		3/3
	Phlebotomine	0/1	0/3
	Tick	9/13	1/5
Orbivirus	Mosquito	1/1	5/5
	Culcicoides		2/2
	Phlebotomine		1/1
	Tick	6/6	6/6
	Unknown		1/1
Rhabdovirus	Mosquito		3/3
	Culicoides		1/1
	Phlebotomine	1/1	3/3
	Tick	2/2	1/1
	Unknown		3/4
Unclassified	Mosquito		2/2
	Tick	10/11	0/10
Arenavirus	None	0/1	0/2
Picornavirus	Mosquito	0/1	2/2

[a]   Known or suspected.

[b]   Ixodid tick cell lines: *Rhipocephalus appendiculatus; Boophilus microplus; Dermacentor paramapertus: D. variabilis* and *Haemaphysalis spinigera.*

[c]   Culicine mosquito cell lines: *Aedes aegypti; A. albopictus; A. pseudoscutellaris; A. malayensis; A. vexans; A. vittatus; A. w-albus; A. dorsalis; C. quinquefasciatus; C. tarsalis; C. molestus; C. tritaeniorhyncus* and *C. inornata.*

270 line from *B. microplus.* ZIR virus, a Hughes group virus isolated from *O. muesebecki,* also showed a long lag phase in the RA-243 cells but not in the BM-270 cells. The reasons for these differences are not evident. It may be that argasid viruses whose exposure to vectors and hosts is limited may be more highly adapted than ixodid viruses which are generally associated with a wide range of vectors and hosts. However, still too few of the aboviruses associated with ticks, particularly argasid ticks, have been tested to draw major conclusions.

Generally, the growth of arboviruses is relatively slow in tick cells and is maintained for long periods without significant loss of titer. No cytopathic effects (CPE) have been observed.

H. Persistent Infections

Generally arboviruses do not cause destructive CPE in arthropod cells although the same viruses cause extensive CPE in vertebrate systems. The cultures are, however, persistently infected, yielding relatively low levels for as long as the cultures are maintained. Many studies have been carried out on the mechanisms of control of this important aspect of arbovirus replication by the use of mosquito cell lines. Similar detailed investigations on this equally important topic in tick cells have yet to be undertaken.

Persistent infection of RA-243 with LI virus has been obtained.[16] The cells were subcultured and subsequently taken through a total of 90 weekly subcultures 28 days p.i. with LI, producing 4 to 6 dex PFU/mℓ/week. Examination of the plaque morphology of the progeny virus has shown no striking evidence for the production of small plaque variants frequently produced in mosquito cell experiments. Persistently infected cells at subculture 40 were superinfected with homologous LI virus, but no amplification in extracellular titer was observed, whereas heterologous SF virus grew to normal levels in persistently infected cells. These results suggest the absence of a broad spectrum interferon-like antiviral defense system in these cells. Similarly persistent infection of BM-270 with LGT virus was also obtained and maintained for 98 days, producing an extracellular virus of between 3 to 5 PFU/mℓ. No alteration in plaque morphology was observed. These are the only persistently infected tick cell lines which have been produced, but it is assumed that as in the case of mosquito cells, persistent infection can readily be obtained provided virus growth has been demonstrated. A wider range of viruses needs to be investigated to confirm that the lack of plaque variation observed in these two examples is a feature of persistence in tick cells.

NSD virus serially passed in RA-243 for 35 passages yielded high titers of virus as measured in mice, but appeared to be attenuated for sheep. The 26th passage material did not produce viremia in a sheep as did the 2nd and 3rd passage material.[22] This is the only example of the use of tick cells in attempts to attenuate viruses for vaccine use, a potentially important area for exploitation of tick cell lines which has already been taken up for mosquito cells particularly with regard to the dengue virus (see Volume II, Chapter 12).

I. Cytopathic Effects

CPE with a limited range of viruses and only under certain conditions has been reported for only a few mosquito cell lines of the many arthropod cell lines used for viral replication. No CPE has ever been reported in any viral replication of tick cell lines. A CPE is useful for direct isolation from field material, for rapid titration and by the use of agar overlay for plaque titrations. With the *Aedes albopictus* and *A. pseudoscutellaris* cell lines it was observed that the CPE could be enhanced by varying the culture conditions (i.e., growing the cells on plastic instead of glass, lowering the osmotic pressure, lowering the pH increasing the incubation temperature, reducing the quantity of fetal calf serum, and generally stressing the cells) and by early passage level cells being required. The passage history of the virus can also greatly influence the cytopathic response: in general the closer to the field isolate and the fewer mouse-brain passages, the greater are the chances of producing CPE in mosquito cell lines.[1] It has also been proven possible to produce clones which respond to certain viruses with a CPE when the parent line did not.[27]

It may be that a tick cell line has not yet been produced from a species suitable for the production of a cytopathic effect, although it is more likely that a suitable cell line has not yet been produced regardless of species of origin. It is also possible that tick cells never respond to viral infection by the production of a CPE which is, in the case of arthropods, an induced in vitro response. The slow growth rate of the cells and some viruses, poor cell/cell contact, and possibly CPE at low levels may all be factors responsible for the lack of a usable CPE. One of the most likely reasons is the way the cell lines have been established. In the majority of cases a long lag-phase has been encountered where heavy selection pressure is apparently taking place and during which period more delicate cells capable of showing viral damage would be lost. With improved culture conditions resulting in the more rapid establishment of cell lines containing a greater variety of cell types, there would be an improved chance of inducing a CPE. Again, partly due to the slow growth rate of the present tick cell lines and the often suboptimal media used, no clones have yet been produced from tick cells. It is possible that a faster growing, healthy tick cell line could be stressed in such a way as to produce a CPE as in the case of mosquito cells. The newer *Boophilus microplus* VIII-633, IX,[28,29] *Rhipicephalus appendiculatus* RAE-25, and the *R. sanguineus* RSE-8[5] and RSE-13[30] has not been used for arbovirus investigations. The possibility exists that mosquito and tick cells might be hybridized and found to grow faster and show a much wider spectrum of arbovirus susceptibility than tick or mosquito cells alone. That these hybrid lines might also respond to infection with tick-borne viruses by production of CPE is not an unreasonable speculation.

III. APPLICATIONS

A. Isolation from Field Material

The lack of CPE production is a limiting factor in extension of the use of tick cells for isolation from field material. However, the sensitivity to infection of both primary cultures and cell lines has been observed, for example RA-243 could be infected with a 10^{-7} dilution of QRF seed virus and RA-14 could be infected with the *Bunyavirus* HAZ at a level too low to detect in Vero cells. Thus, tick cell lines could be used to "boost" low infectivity of field collected material to levels easily detected in other systems.

The recent isolation of a new serologically unique, filterable agent Cascade virus from *Dermacentor occidentalis* ticks in North America[31] emphasizes the importance of appropriate isolation systems for detecting presumably tick related pathogens. The Cascade virus could easily be isolated in the poikilothermic amphibian *Xenopus laevis* (XTC-2) cell line[20] grown at 27 or 34°C in which it produced CPE and plaques, or in the *D. variabilis* (RML-15) cell line grown at 27°C, but not in laboratory animals nor vertebrate cell cultures. This virus would never have been isolated had only conventional mammalian systems been used. The material isolated directly from ticks now provides a unique example of this virus which has never been passaged in vertebrate cells. The fact that the viral envelope is derived from the host cell membrane underlines the importance of this isolation, as tick-propagated virus may have altered infectivity for heterologous host cells and vice versa. Similarly, DEN-2 (strain PR-159) isolated in AP-61 mosquito cells also could not be detected in PS cells and had a titer 3 dex lower in suckling mice than by CPE in AP-61.[1] The AP-61 cell line has been used for the direct isolation of the DEN virus from field material in the Caribbean. The hardiness of this particular cell line and its ability to grow well at ambient temperatures makes it ideal for this type of field work, and the inoculation of samples directly into cells where multiplication can take place during transportation increases the chance of viral isolation. Although tick cell lines are generally not as hardy as the AP-61 cell line, direct inoculation of sera or tick pools could be used to detect very low levels of virus with further titration in XTC-2 as used for Cascade virus.

B. Immunofluorescence and Immunocytochemistry

Isolation from field material when known viruses are suspected can be carried out using an indirect fluorescent antibody method on the tick cells as is now widely used for mosquito-borne viruses in mosquito cells. To test the feasability of this approach and to further test the range of viruses that would replicate in RA-243, 12 viruses were inoculated separately and after 8 days growth was determined by an indirect immunofluorescence test employing specific immune ascitic fluid in serial dilutions in chamber slides. Five of the viruses, POW, QYB, UUK, TRB, and SAW were detected in this way. All but QYB are ixodid tick-transmitted. Of the negative results, Pichinde is not strictly an arbovirus (it is a *Rhabdovirus* of the LCM-Tacaribe group) and CNU and GA are argasid viruses which may not have reached sufficiently high titers by 8 days. All material was taken directly from suckling mouse brain material and in 2 cases also failed to grow in Vero cells. As growth kinetics of individual viruses are determined, this method can be used with greater accuracy. This method was also used in a large scale attempt at isolating the causative agent of Lyme disease from its suspected vector *Ixodes dammini.* Large numbers of ticks were collected, triturated, and added to the RA-243 cells which were treated at various time intervals by the indirect fluorescent antibody method using convalescent serum. Although no agent was detected, this may have been due to the presence of antibiotics, as the etiological agent of Lyme disease is now thought to be a spirochete.[32]

This is another area which, although widely exploited for mosquito-cells and viruses, has not been applied to tick cells, although it overcomes the drawback of lack of CPE production and exploits the sensitivity of these cells for tick-borne viruses. The advent of monoclonal antibodies to mosquito-borne viruses such as DEN, JE, YF, and of the *Flavivirus* group-specific monoclonals have provided new tools with which mosquito cells may be used for rapid diagnosis and typing of viruses by indirect immunofluorescence or immunoperoxidase staining. Likewise, tick cells in combination with enzyme-linked immunosorbent assay (ELISA) could be used to detect very low levels of viral antigen, as has been shown for mosquito cells.

C. Contamination

Evidence for virus-like contamination of mosquito cell lines is widespread[1] and the relationship with the production of CPE is uncertain. In the case of AP-61, there is strong evidence to suggest that the contaminating picornavirus, Kawino, came from the original mosquito larvae used to initiate the cell lines,[1] as did the rod-shaped virus found in the *Anopheles stephensi* cell lines.[33] The widely distributed ATC-15 cell line was at one time shown to contain 5 types of virus-like particles although the Igarashi clone C6/36 is considered to be virus free. There is only one report of virus-like particles in one *Rhipicephalus appendiculatus* cell line[34] and Munz has verified this (see Chapter 11). However, as most of these particles are observed when large scale electron microscope or molecular characterization of viruses is carried out, this is more likely due to the lack of sufficient observations rather than lack of viruses. With more detailed investigations it is likely that additional virus-like particles particularly of tick origin will be identified and characterized.

IV. CONCLUSION AND SUMMARY

A wide range of arboviruses, both mosquito and tick-borne, will replicate in tick cells. No cytopathic effect is produced. In general, viruses isolated from ixodid ticks multiply more readily and to a higher titer than do those from argasid ticks, although QRF virus grew readily in RA-243, but not in RM-14.

Several viruses showed long-lag phases before an extracellular virus could be detected, particularly the argasid transmitted Hughes group viruses in *Dermacentor* cells. Although ZIR virus had a considerable lag phase in the RA-243, it did not in the BM-270. Unfortunately still too few viruses of each group have been tested to draw overall conclusions. In this case the differences could have been due to innate characteristics of specific cell lines (i.e., RA-243 is a much older line than either RM-14 and BM-270) or variation in strain and passage history of virus used plus variations in inoculum.

The majority of tick-borne viruses tested grew in one or more of the cell lines used. The negative results usually occurred where only one cell line was used or one method of detection (i.e., immunofluorescence) in which optimum conditions for individual viruses were not determined, was employed. Therefore it should be possible to find a tick cell line for growth studies of most tick-borne arboviruses. In combination with mosquito cell lines the tick cell lines can be used to further characterize new isolates in as much as tick-borne viruses, except orbiviruses generally do not grow in mosquito cells and nonarboviruses, apart from arthropod viruses which do not grow in either of these cell types.

The further development of the use of tick cell lines has not followed that of the mosquito cell lines. This is mainly because of the slow growth rate of tick cells and the frequent difficulty obtained in transporting them from one laboratory to another, although the original RA-243 cell line has now been in continuous cultivation for 12 years.

The discovery that tick egg extract could stimulate cell growth and that juvenile hormone had a bimodal effect[30] indicates that more detailed studies on growth factors and better media could lead to improvements in the production of tick cell lines. This could result in a greater variety of cell types and easier maintenance of cell lines with the possibility of a broader spectrum of use. Clones might also be produced that possess increased sensitivity to certain viruses or the ability to produce a CPE.

The work reviewed here provides a sound baseline of information for more detailed future work on individual viruses and tick cell combinations of particular interest. There is no reason why the numerous applications for which the mosquito cell lines have proved useful, as witnessed by this book, should not also be applied to tick cell lines. The lines are available and much basic information has been accumulated; what is currently missing are investigators in this important area of research, an area whose true potential may never be realized. Of particular relevance is the use of these systems for the isolation of pathogens from field material where the use of nonmammalian cell lines may detect unique strains previously undetected. Current work with tick cell cultures is more concerned with studies of tick-borne parasites than with arboviruses. It is hoped that improvements in culture technique might result from this approach and will lead to a revitalized interest in tick cell lines. These lines are at risk of becoming endangered species after such a long fight for existence.

ACKNOWLEDGMENTS

Much of the work reviewed here was carried out under successive grants from the Medical Research Council of Great Britain in collaboration with my colleagues Prof. M. G. R. Varma and Dr. C. J. Leake. I am presently a member of the Department of Biochemical Microbiology of the Wellcome Research Laboratories and am grateful for their support during the preparation of this manuscript.

REFERENCES

1. **Pudney, M., Leake, C. J., and Buckley, S. M.,** Replication of arboviruses in arthropod *in vitro* systems: an overview, in *Invertebrate Cell Culture Applications,* Maramorosch, K. and Mitsuhashi, J., Eds., Academic Press, New York, 1982, 159.
2. **Weyer, F.,** Explanationsversuche bei Läusen in Verbindung mit der Kultur von Rickettsien, *Zentralbl. Bakteriol. Parasitenkd. Infektionskr. Hyg. Abt.,* 159, 13, 1952.
3. **Varma, M. G. R., Pudney, M., and Leake, C. J.,** The establishment of three cell lines from the tick *Rhipicephalus appendiculatus* (Acari: Ixodidae) and their infection with some arboviruses, *J. Med. Entomol.,* 11, 698, 1975.
4. **Kurtti, T. J. and Buscher, G.,** Trends in tick cell culture, in *Practical Tissue Culture Applications,* Maramorosch, K. and Hirumi, H., Eds., Academic Press, New York, 1979, 351.

5. Kurtti, T. J. and Munderloh, U. G., Tick cell culture: characteristics, growth requirements, and applications to parasitology in *Invertebrate Cell Culture,* Maramorosch, K. and Mitsuhashi, J., Eds., Academic Press, New York, 1982, 195.

6. Pudney, M., Varma, M. G. R., and Leake, C. J., Establishment of cell lines from Ixodid ticks, *TCA Manual,* 5, 1003, 1979.

7. Yunker, C. E. and Meibos, H., Culture of embryonic tick cells (Acari: Ixodidae), *TCA Manual,* 5, 1015, 1979.

8. Rehacek, J., Cultivation of different viruses in tick tissue culture, *Acta Virol. Engl. Ed.,* 99, 322, 1965.

9. Pudney, M., Cell Culture of Tissues from Arthropods and Lower Vertebrates for Application in the Study of Arboviruses, Unpublished Ph.D. thesis, Univeristy of London, 1973.

10. Yunker, C. E. and Cory, J., Growth of Colorado tick fever virus in primary tissue cultures of its vector *Dermacentor andersoni* Stiles (Acarina: Ixodidae), with notes on tick tissue culture, *Exp. Parasitol.,* 20, 267, 1967.

11. Rehacek, J., Use of invertebrate cell culture for study of animal viruses and rickettsiae, in *Invertebrate Tissue Culture,* Vol. 2, Vago, C., Ed., Academic Press, New York, 1972, 279.

12. Bhat, U. K. M. and Yunker, C. E., Susceptiblity of tick cell line (*Dermacentor parumapertus* Neumann) to infection with arboviruses, in *Arctic and Tropical Arboviruses,* Kurstak, E., Ed., Academic Press, New York, 1979, 263.

13. Yunker, C. E., Cory, J., and Meibos, H., Continuous cell lines from embryonic tissues of ticks (Acari: Ixodidae), *In Vitro,* 17, 139, 1981.

14. Guru, P. Y., Dhanda, V., and Gupta, N. P., Cell cultures derived from the developing adults of three species of ticks by a simplified technique, *Indian J. Med. Res.,* 64, 1041, 1976.

15. Pudney, M., Varma, M. G. R., and Leake, C. J., Culture of embryonic cells from the tick *Boophilus microplus* (Ixodidae), *J. Med. Entomol.,* 10, 493, 1973.

16. Leake, C. J., Pudney, M., and Varma, M. G. R., Studies on arboviruses in established tick cell lines, in *Invertebrate Systems In Vitro,* Kurstak, E., Maramorosch, K., and Dübendorfer, A., Eds., Elsevier/North Holland, Amsterdam, 1980, 327.

17. Pudney, M., Leake, C. J., and Varma, M. G. R., Replication of arboviruses in arthropod *in vitro* systems, in *Arctic and Tropical Arboviruses,* Kurstak, E., Ed., Academic Press, New York, 1979, 245.

18. Pudney, M., Varma, M. G. R., and Leake, C. J., The growth of some arboviruses in tick cell lines, in *Proc. Int. Conf. Tick-borne Diseases and Their Vectors,* Edinburgh, 1976, 490.

19. Leake, C. J., Comparative growth of arboviruses in cell lines derived from *Aedes* and *Anopheles* mosquitoes and from the tick *Boophilus microplus,* in *Arboviruses in Arthropod Cells In Vitro,* CRC Press, Boca Raton, Fla., 1987, chap. 13.

20. Pudney, M., Varma, M. G. R., and Leake, C. J., Establishment of a cell line (XTC-2) from the South African clawed toad, *Xenopus laevis, Experientia,* 29, 466, 1973.

21. Leake, C. J., Varma, M. G. R., and Pudney, M., Cytopathic effect and plaque formation by some arboviruses in a continuous cell line (XTC-2) from the toad *Xenopus laevis, J. Gen. Virol.,* 35, 335, 1977.

22. Munz, E., Reimann, M., Munderloh, U., and Settel, U., The susceptibility of the tick cell line TTC-243 for Nairobi sheep disease virus and some other important species of mammalian RNA and DNA viruses, in *Invertebrate Systems In Vitro,* Kurstak, E., Maramorosch, K., and Dübendorfer, A., Eds., Elsevier/North Holland, Amsterdam, 1980, 337.

23. Chen, Z. S. and Buckley, S. M., Personal communication quoted in Reference 1, 1982.

24. Pudney, M., Steere, A. C., and Grodzicki, R., Arbovirus growth in tick cells using immunofluorescence, Unpublished results, 1981.

25. Rehacek, J., Cultivation of different viruses in tick tissue culture, *Acta Virol. Engl. Ed.,* 9, 332, 1965.

26. Pudney, M., Newman, J. F. E., and Brown, F., Characterisation of Kawino virus, an entero-like virus isolated from the mosquito *Mansonia uniformis* (Diptera: Culicidae), *J. Gen. Virol.,* 40, 433, 1978.

27. Sarver, N. and Stollar, V., Sindbis virus-induced cytopathic effect in clones of *Aedes albopictus* (Singh) cells, *Virology,* 80, 390, 1977.

28. Holman, P. J. and Ronald, N. C., A new tick cell line derived from *Boophilus microplus, Res. Vet. Sci.,* 29, 383, 1980.

29. Holman, P. J., Partial characterization of a unique female diploid cell strain from the tick *Boophilus microplus* (Acari: Ixodidae), *J. Med. Entomol.,* 18, 84, 1981.

30. Kurtti, T. J., Munderloh, U. G., and Samish, M., Effect of medium supplements on tick cells in culture, *J. Parasitol.,* 68, 930, 1982.

31. Yunker, C. E., Cory, J., Gresbrink, R. A., Thomas, L. A., and Clifford, C. M., Tickborne viruses in western North America. III. Viruses from man-biting ticks (Acari: Ixodidae) in Oregon, *J. Med. Entomol.,* 18, 457, 1981.

32. Burgdorfer, W., Barbour, A. G., Hayes, S. F., Benach, J. L., Grunwaldt, E., and Davis, J. P., Lyme disease — a tick-borne spirochetosis?, *Science,* 216, 1317, 1982.
33. Pudney, M., McCarthy, D., and Shortridge, K. F., Rod-shaped virus-like particles in cultured *Anopheles* cells and in an *Anopheles* laboratory colony, in *Proc. Third Int. Colloq. on Invertebrate Tissue Culture,* Smolenice, Czechoslovakia, 1973, 337.
34. Williams, H., Personal communication quoted in Reference number 1.
35. Banerjee, K., Guru, P. Y., and Dhanda, V., Growth of arboviruses in cell cultures derived from the tick *Haemophysalis spinigera, Indian J. Med. Res.,* 66, 530, 1977.

Chapter 8

APPLICATION OF *AEDES ALBOPICTUS* CLONE C6/36 CELLS TO THE ISOLATION OF MOSQUITO-BORNE TOGAVIRUSES IN JAPAN, INDONESIA, AND THAILAND

A. Igarashi

TABLE OF CONTENTS

I. INTRODUCTION

Among arthropod-borne vertebrate viruses (arboviruses) several mosquito-borne togaviruses have been associated with severe human diseases such as encephalitis and hemorrhagic fevers. Isolation of these viruses from clinical and field materials is important for laboratory diagnosis as well as for epidemiological surveys of the viruses in nature. The procedures have commonly been performed by inoculating test specimens into brains of suckling mice (SMB).[1] This method is sensitive enough to detect neurotropic viruses, especially those causing encephalitis. However it is not always efficient in the isolation of viruses or mutants of viruses having less neurotropism than the standard virus. A generation of such mutant viruses in cultured mosquito cells persistently infected with several togaviruses has been well documented.[2-5] Also there is evidence that some "mutant" viruses appear in laboratory colonized virus-infected mosquitoes or other insects.[6,7] These mutants appear to be more easily isolated in cultured mosquito cells rather than by inoculation of SMB or other mammalian hosts.[8] Therefore attempts were made to test cultured mosquito cells as alternative systems for the isolation of mosquito-borne togaviruses.

Before testing field materials we attempted to set up a sensitive host cell system for the isolation of dengue viruses because the viruses have been associated with a large numbers of infections in Southeast Asia, which sometimes results in severe manifestations of hemorrhagic and shock syndrome.[9,10] Sensitivity to the chikungunya virus was also tested because of its occasional association with the disease in children in the same epidemic season. *Aedes albopictus* cells originally established by Singh[11] were cloned and the clones were selected for virus supporting capacity. The parent cell line had first been adapted to grow in a medium which consisted of 90% of the medium of Eagle[12] and 10% of the medium of Mitsuhashi-Maramorosch (MM)[13] supplemented with 0.2 mM each of nonessential amino acids, and finally without MM medium. By repeated cloning and testing of each clone for its capacity to produce each of the 4 types of dengue and chikungunya viruses, clone C6/36 was selected as a suitable one for further studies.[14] Results of the application of C6/36 cells to the isolation of mosquito-borne togaviruses in Japan by our group are described below along with some experiences in Indonesia and Thailand.

II. COMPARATIVE SENSITIVITY OF C6/36 CELLS AND SUCKLING MOUSE BRAIN (SMB) INOCULATION FOR THE ISOLATION OF JAPANESE ENCEPHALITIS (JE) AND GETAH VIRUSES IN JAPAN

A. Isolation of JE Virus From Field-Caught Mosquitoes

The JE virus was first isolated in Japan in 1935, hence its name. However the virus is known to exist from East-, Southeast-, to South Asia causing severe encephalitis.[15] In Japan the disease occurs every year in late summer to early autumn and is caused by bites of infected vector mosquitoes, *Culex tritaeniorhynchus*. Since 1966 the number of apparent encephalitis cases in Japan has decreased year by year, and the annual reported cases has never exceeded 100 during the past 10 years. However, the virus circulates in nature every summer between vector mosquitoes and swine; the latter is the most important amplifier of the virus. The activity of the virus in nature has been monitored either by isolation of the virus from vector mosquitoes or by antibody surveys of slaughtered swine. The former is performed by inoculating mosquito homogenates into SMB, observing neurological symptoms, and detecting virus antigens in the brains of moribund mice either by the fluorescent antibody method or by the hemagglutination-inhibition (HI) test after extracting the antigen.[1]

The author and his co-workers tested C6/36 cells as a substitute for SMB for the

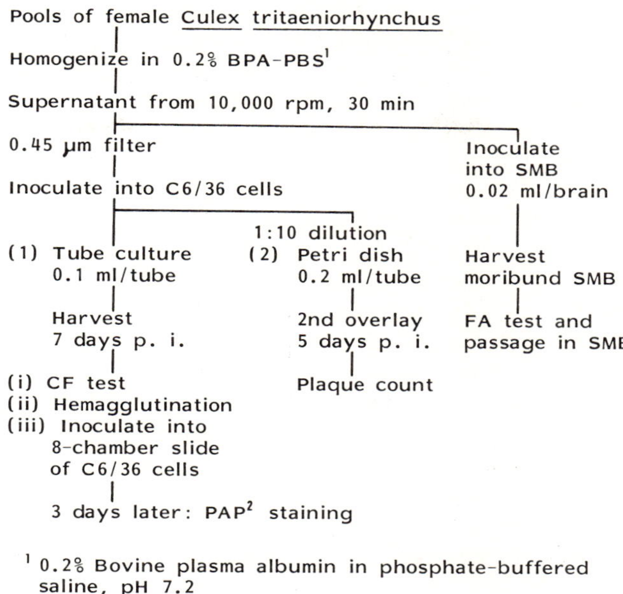

FIGURE 1.   Virus isolation from field-caught *Culex tritaeniorhynchus* in *Aedes albopictus* clone C6/36 cells and by SMB inoculations.

isolation of the virus, hoping that the method would provide a new and easier tool for field studies and also make it possible to detect viruses or mutants of viruses that can be isolated only by C6/36 cells. This study began in 1978 and continued for 5 years. It was performed for the first 2 years in Osaka and thereafter in Nagasaki in collaboration with the Osaka Prefectural Institute of Public Health and the Nagasaki Prefectural Institute of Public Health and Environmental Sciences, respectively.

The isolation procedure is illustrated in Figure 1.[16] Field-collected *C. tritaeniorhynchus* females were pooled (approximately 100 mosquitoes/pool), homogenized, and centrifuged. The resulting supernatant was inoculated in SMB and the remaining was filtered through 0.45 μm filters. The filtrate was inoculated into tube cultures of C6/36 cells and sometimes also to petri dish cultures of C6/36 cells in order to produce plaques under agar overlay. In 1978 three methods were used to detect infective virus or virus antigens in culture fluids of C6/36 cells: (1) complement-fixation (CF) test using standard anti-JE serum, (2) detection of hemagglutinin by goose red blood cells, (3) inoculation of the infected culture fluid to C6/36 cells on eight-chamber slides followed by immunoperoxidase (PAP) staining of intracellular viral antigens.[17] Out of 29 strains of JE virus detected by the PAP method, 28 were positive by the CF. PAP staining was used thereafter in our laboratory. However, the CF test is also easy to perform when facilities and experience are available.

Comparative isolation of JE virus in C6/36 and SMB during the 5 years of study is summarized in Table 1 which shows higher efficiency of C6/36 compared with SMB for JE virus isolation from field-caught mosquitoes. In 1978, 3 specimens did not yield JE virus in C6/36 cells, although they did in SMB. Two of the three pools had been kept frozen after SMB inoculation before inoculation in C6/36 cells; thus the virus was possibly inactivated during freezing and thawing as is sometimes observed for JE virus. Another specimen produced an agent(s) in C6/36 cells that inhibited the growth of JE virus. One specimen collected in 1980 was negative for JE virus in C6/36, although the

Table 1

ISOLATION OF JE AND GETAH VIRUSES FROM FIELD-CAUGHT
CULEX TRITAENIORHYNCHUS IN *AEDES ALBOPICTUS* CLONE C6/
36 CELLS AND BY SMB[a]

		Osaka		Nagasaki			
Virus	Isolation by	1978	1979	1980	1981	1982	Total of 5 years (%)
JE	C6/36	43	51	17	23	14	148 (11.7)
	SMB	27	50	8	20	9	114 (9.0)
Getah	C6/36	2	1	3	0	0	6 (0.48)
	SMB	0	0	2	0	0	2 (0.16)
Number of pools tested		349	342	180	190	200	1261 (100)

[a]   Figures represent the number of mosquito pools.

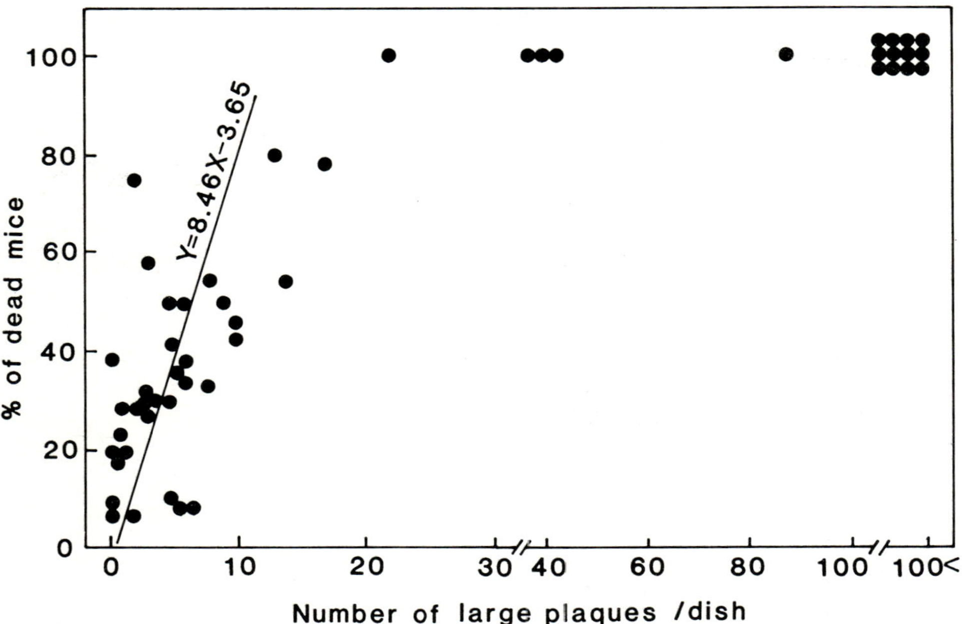

FIGURE 2.   Comparison between the number of plaques on C6/36 cells and mortality of suckling mice inoculated with field-caught mosquito homogenate which yielded JE virus. Each dot represents each pool of *Culex tritaeniorhynchus* with positive JE virus isolation. Each mosquito pool was processed as described in the text and Figure 1 and inoculated into a litter of 8 SMB (0.02 m*l*/brain). The remainder was membrane filtered and 1:10 diluted filtrate was inoculated into C6/36 cells on Petri dishes (0.2 m*l*/dish) to form plaques. Percentage of dead mice in the litter was scored on the ordinate and the number of plaques resembling those of standard JE virus on the abscissa.

specimen yielded JE virus in SMB. This result was probably due to bacterial contamination of cell cultures. In 1979 each mosquito homogenate was inoculated into a litter of SMB and, after filtration, into C6/36 cells both in tube cultures and in Petri dishes in order to produce plaques. Figure 2 compares the number of plaques (X) resembling those of standard JE virus and the percentage of mice dead (Y) after inoculation of the same mosquito homogenate, as observed for each of the 50 pools which gave JE virus isolates. The result shows that when X is over 20 (that is, more than 20 plaque-forming

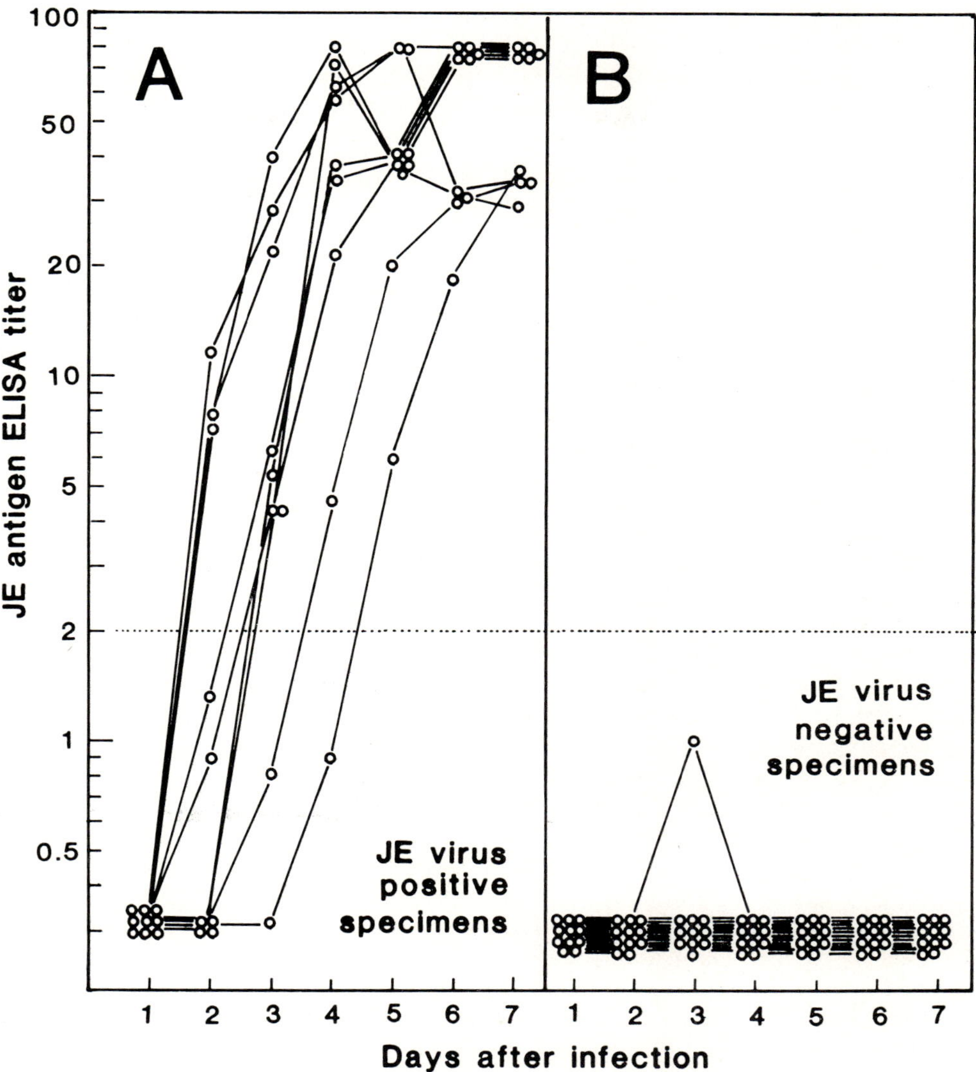

FIGURE 3.   Production of JE-ELISA antigen in C6/36 cell culture fluid after inoculation of field-caught *Culex tritaeniorhynchus* homogenate. Tube cultures of C6/36 cells were inoculated with the mosquito homogenates and incubated at 28°C under maintenance medium. Infected culture fluid was examined daily for the production of JE-ELISA antigen by the sandwich method. Results with 9 specimens which yielded JE virus (A), and those with 11 JE negative specimens (B) are shown.

units of the virus were inoculated) Y was 100 (that is, all the inoculated mice died). For the specimens with X less than 20, the correlation coefficient between X and Y was 0.6, with the equation of linear regression: $Y = 8.46X - 3.66$. Following this equation, for the value of $Y = 50$, X was estimated approximately as 6. In other words, when 6 plaque-forming units of virus were inoculated into SMB, the probability is 50% that the mouse will be killed, indicating an approximately 10 times higher sensitivity of C6/36 cells compared with SMB when one LD_{50} was considered as 0.6 infectious units.

In 1982, detection of a JE virus antigen was tested by the enzyme-linked immunosorbent assay (ELISA).[18] Daily after inoculation of field-caught mosquito homogenates, C6/36 cell culture fluids were examined by the sandwich ELISA. As shown in Figure 3, in the cultures that eventually turned out to be positive for JE virus as shown by the immunoperoxidase method, the virus antigen became detectable from 2 to 5 days after the inoculation. On the other hand, the cultures that did not yield JE virus never produced a detectable amount of virus antigen. The infected C6/36 cell culture fluids harvested 7 days after the inoculation of test materials during the past 4-year study had been kept at −70°C. A total of 256 specimens (110 JE positives and 146 JE negatives) was examined and the result by the ELISA completely agreed with the result of previous virus detection by the immunoperoxidase method.[19]

B. Isolation of Getah Virus From Field-Caught Mosquitoes

Epidemiological surveys of JE virus sometimes pick up non-JE arboviruses and the Getah virus, an Alphavirus of the family Togaviridae, is among these. Recently, the Getah virus has drawn attention because of its economic impact. It was shown to cause infections associated with respiratory symptoms and eruptions among race horses.[20] During our virus isolation studies, the Getah virus was included in the screening by using standard anti-Getah virus serum. Resulting Getah virus isolation rates again show higher sensitivity of C6/36 cells compared with SMB inoculation (Table 1).

C. Isolation of JE Virus From Post-Mortem Brain Material

Laboratory diagnosis of JE consist of virus isolation and serology.[1] The former is more reliable but can be used only for fatal cases which decreased dramatically in the past 10 years in Japan. In 1980 we had a chance to examine a case of fatal encephalitis in a 61 year-old male who developed acute encephalitis on August 26 and died 4 days later. Personnel of the Nagasaki Prefectural Institute of Public Health and Environmental Sciences attempted without success to isolate the virus from post-mortem brain materials by SMB inoculation. The remaining materials were sent to our laboratory and inoculated into C6/36 cell cultures, yielding JE virus from cerebrum, cerebellum, and pons; however, cerebrospinal fluid did not yield JE virus.[21]

III. ISOLATION OF JE VIRUS FROM SLAUGHTERED SWINE

In order to monitor JE virus in nature, antibody surveys among slaughtered swine are routinely performed in Japan every summer. In contrast, virus isolation from the same material is not so widely performed because of technical and economical reasons. In 1981 we used C6/36 cells to isolate JE and Getah viruses from heparinized swine blood collected at a slaughter house in Nagasaki Prefecture. The specimens (B) were separated into lymphocyte (L) and plasma-platelet (P) fractions by centrifugation through Ficoll-Pack. Each of the B, L, and P fractions was inoculated to tube cultures of C6/36 cells and the presence of JE and Getah viruses was screened. Serum antibody titers against JE virus were tested by the HI test. Results (Figure 4) show the mosquito-swine-mosquito amplification of JE virus in nature. In this study we did not perform comparative sensitivity tests of C6/36 cells against SMB inoculation, however, results demonstrate applicability of C6/36 cells to epidemiological studies.[22]

IV. ISOLATION OF MUTANT VIRUSES FROM FIELD MATERIALS USING C6/36 CELLS

As previously mentioned, one of the purposes of using C6/36 cells was to detect ''mutant'' viruses that are hard to isolate in SMB and to demonstrate the production of such ''mutants'' in nature, as has been shown by laboratory experiments. All the JE and Getah virus isolates obtained during the 5 year study in Osaka and Nagasaki were reviewed as to whether they were isolated in C6/36 cells or in SMB. Those isolated only in C6/36 cells were carefully examined for their plating efficiencies (EOP) on

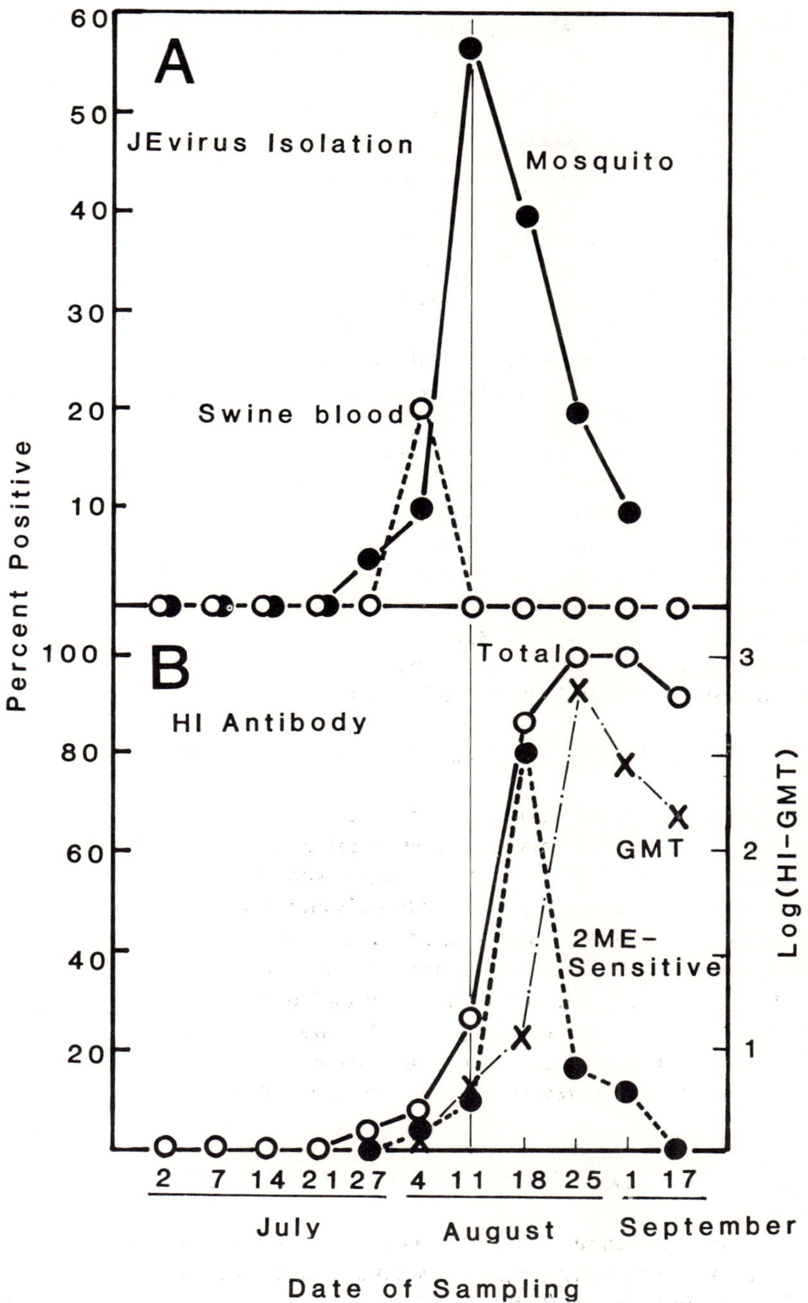

FIGURE 4.   Isolation of JE virus by C6/36 cells from field-caught *Culex tritaenior-hynchus* and blood of slaughtered swine and production of anti-JE antibodies in sera of slaughtered swine. In the epidemic season in Nagasaki, 1980, mosquito homogenates and swine bloods collected weekly were inoculated into C6/36 cells to isolate JE virus (A). Anti-JE antibodies in slaughtered swine sera were examined by the HI test combined with 2-mercaptoethanol (2-ME) treatment to detect IgM antibodies (B). Antibody titers are shown by geometrical mean titer (GMT).

C6/36 cells at 28°C and on BHK21 cells at 37°C. None of the isolates clearly showed significantly higher EOP on C6/36 cells compared with BHK21 cells. Thus, evidence was not obtained showing that such mutant virus strains arose in infected mosquitoes in nature. In 1978 two strains of Getah viruses (B42 and CO1) were isolated only in C6/36 cells. These strains were inoculated in C6/36 cells in petri dishes to produce plaques and each plaque progeny, after enrichment in C6/36 cells under liquid media, was tested for its EOP on C6/36 cells at 28°C and on BHK21 cells at 37°C. Compared with BHK21 cells or LD_{50} in SMB inoculation, 2 of the 23 plaque isolates from B42 (P14 and P15) showed significantly higher EOP on C6/36 cells. Thus, these plaque isolates were considered as mutant viruses. Similar plaque mutants were again isolated in 1978 by direct plaque formation on C6/36 cells from the original homogenate of mosquito pool B49. However, no such mutant viruses were isolated from another Getah virus isolate, C01, or the standard strains of Getah virus, (AMM2021, Sagiyama) which were isolated by SMB inoculation. The proportion of mutant viruses in the B42 isolate did not change significantly after ten serial passages in C6/36 cells; however it decreased to an undetectable level by a single passage in SMB.[23] In Nagasaki in 1980, Getah virus was isolated from the same mosquito pool both in C6/36 cells and by SMB inoculation. The former strain (12380A) yielded two mutant plaques out of the 24 plaques tested. However, no such mutant plaques were observed in the latter strain (12380M). These mutant plaque isolates differed from their parental strains in their RNA fingerprints. Also the fingerprint of 12380A was different from that of 12380M.[24] By direct plaque isolation, a mutant of JE virus was isolated out of 237 plaques formed by 34 mosquito pools collected in 1979.[23]

V. ISOLATION OF INSECT VIRUSES IN C6/36 CELLS

When the homogenates of field-caught mosquitoes were inoculated into C6/36 cells and covered by agar overlay, many plaque-forming agents were detected.[16] Some of these agents were JE or Getah viruses, but many were unrelated to these arboviruses. Until now only three of the plaque isolates have been examined in some detail in Japan and all of them appeared to possess common features, indicating that they may be considered as insect viruses of mosquitoes.[25-27] They do not multiply in mammalian hosts or cell cultures, although they can grow and form plaques in C6/36 cells and some can infect mosquito larvae. Morphological and biophysical examination appears to show that these insect viruses do not comprise a uniform entity. A similar plaque-forming agent has been isolated from leafhoppers which were captured on East China Sea.[28]

VI. ISOLATION OF DENGUE VIRUSES IN C6/36 CELLS IN INDONESIA

Infections caused by dengue viruses are widespread in Southeast Asia, and Indonesia is not an exception. Several studies have been made on dengue and dengue hemorrhagic fever (DHF) in Indonesia. In 1980 to 1982, Japanese study teams visited Jakarta and performed virus isolations as a part of the virological, hematological, and immunological studies on DHF in Indonesia. In 1981, 10 cases of DHF and 99 cases of fever of unknown origin (FUO) were examined for virus. Heparinized bloods (B) were separated into lymphocyte (L) and plasma-platelet (P) fractions, as described in Section III, before inoculation in C6/36 cells. Dengue virus isolates were screened by anti-JE rabbit serum (a broadly reactive flavivirus antibody) to detect intracellular virus antigens by PAP staining in C6/36 or BHK21 cells inoculated with infected culture fluids. The positive isolates were identified by PAP staining using type-specific dengue monoclones.[29] Virus isolation was negative for DHF, however, 20 strains of dengue viruses

Table 2
ISOLATION OF DENGUE VIRUSES BY C6/36 CELLS IN JAKARTA, INDONESIA

Year Source		1981		1982			
		DHF[a]	FUO[b]	DHF	FUO	*Aedes aegypti*	*A. albopictus*
Number tested		10	99	15	57	8[c]	28[c]
Virus[d]	D1		2			2	
isolated	D2		6		1		
	D3		5	2	2		

[a]   Dengue hemorrhagic fever.
[b]   Fever of unknown origin.
[c]   Number of mosquito pools.
[d]   D1, D2, D3: dengue virus type 1, type 2, and type 3, respectively.

were isolated from FUO cases; 14 of them were typed in Japan: 2 as type 1, 7 as type 2, and 5 as type 3 viruses, respectively. The data (Table 2) show that three different types of dengue viruses were circulating in a relatively limited study area in a short period. The study also showed that fractionation of heparinized blood into L and P fractions did not improve virus isolation. Thus the isolation in 1982 was done only with unfractionated heparinized blood (B). In that year, several pools of *Aedes albopictus* and *A. aegypti* were also tested together with 15 DHF and 57 FUO cases. The results (Table 2) again showed three types of dengue viruses circulating in one study area during a short period. In that year in addition to C6/36 cells, *Toxorhynchites amboinensis*, TRA171,[30] and human monocytic leukemia J-111[31] cells were also used. The result by TRA171 was the same as C6/36, while J-111 could not detect any dengue virus.[32]

VII. ISOLATION OF DENGUE AND JE VIRUSES IN C6/36 CELLS IN THAILAND

Among Southeast Asian countries, Thailand has recorded the largest numbers of hospitalizations of DHF. Many excellent clinical, serological, pathophysiological, and epidemiological studies on DHF have been reported by Thai investigators and the SEATO (later AFRIMS) laboratory. As shown by Tesh,[33] C6/36 cells possess sensitivity similar to that of mosquito inoculations[34] for the isolation of dengue viruses from the sera of a patient. Fukunaga et al.[35] and Rojanasuphot et al.[36] utilized C6/36 cells for dengue virus isolation; however they relied on the appearance of cytopathic effect (CPE) to detect virus isolates. In our experience, although not all dengue virus isolates produce significant degrees of CPE, dengue virus antigen may be demonstrated in the infected C6/36 cells. Thus, the method of screening described in the preceding section was used in our studies in Chiang Mai in 1982.[37] Virus isolations were performed as a part of the virological and epidemiological studies of encephalitis in this area. Results of tests on the sera of 177 hospitalized patients, 3 post-mortem brains, and certain other specimens are summarized in Table 3. It is noteworthy that each one of the dengue virus types 1 and 2 was isolated from sera of encephalitis and aseptic meningitis patients, respectively, in addition to nine strains from DHF cases. JE virus was isolated from one of the post-mortem brains. Compared with the number of apparent JE cases, isolation of JE virus from field-caught *Culex* mosquitoes was negative, although 2 strains of unidentified flaviviruses and 59 strains of unknown, filterable, CPE-inducing agents were isolated.

Table 3
ISOLATION OF VIRUSES IN C6/36 CELLS IN CHIANG MAI, THAILAND, 1982

Specimens		Mosquito pool[a]	Patients			Normal human sera	HI(−) Swine sera
			Sera[b]	Brain	Liver		
Number		125	177	3	1	50	23
	JE			1			
	D1		8				
	D2		2				
Virus	D3		1				
	Flavivirus	2					
	Unknown	59					

[a]   *Culex tritaeniorhynchus, C. gelidus,* and *C. fuscocephala.*
[b]   Includes 55 encephalitis, 79 dengue hemorrhagic fever, 12 fever of unknown origin, and 8 meningitis.

Modified from Igarashi, A., et al., *Southeast Asian J. Trop. Med. Public Health*, 14, 470, 1983.

VIII. CONCLUSION

The advantages of C6/36 cells for the isolation of JE and Getah viruses are as follows: isolation may be performed with less cost, space, manpower, and time and with similar or even better sensitivity as compared to SMB inoculation. These cells can detect mutant viruses, and many unknown filterable agents which cannot be detected by SMB. However, we must also consider their disadvantages; i.e., the need for facilities and techniques of cell culture, including more strict sterile conditions than SMB inoculation, such as filtration of test materials. One of the merits of C6/36 cells is their capacity to grow at ambient temperatures in tropical areas so that no special incubators are necessary. Our virus isolations in Chiang Mai were performed under this condition, as were the identification and typing of the isolates by monoclones. Since 1980, the Osaka Prefectural Institute of Public Health has dicontinued the use of SMB inoculation, relying instead on C6/36 cells for JE virus isolation. This method is also used in China for JE virus isolation.[38]

It appears important to check the sensitivity of the cell lines before using them for virus isolations, for example, by immunofluorescent or immunoperoxidase staining of the cells infected with standard viruses. It is noteworthy that some latent viruses may contaminate mosquito cell lines.[39,40] When the virus is the same species to be studied, the cell lines will be refractory to the virus because of resistance to superinfection.

REFERENCES

1. Shope, R. E. and Sather, G. E., Arboviruses, in *Diagnostic Procedures for Viral, Rickettsial and Chlamydial Infections,* 5th ed., Lennette, E. L. and Schmidt, N. I., Eds., American Public Health Association, Washington, D. C., 1979, 767.
2. Shenk, T. E., Koshelnyk, K. A., and Stollar, V., Temperature-sensitive virus from *Aedes albopictus* cells chronically infected with Sindbis virus, *J. Virol.*, 13, 439, 1974.
3. Stollar, V., Peleg, J., and Shenk, T. E., Temperature-senstivity of a Sindbis virus mutant isolated from persistently infected *Aedes aegypti* cell cultures, *Intervirology,* 2, 337, 1974.

4. Igarashi, A., Koo, R., and Stollar, V., Evolution and properties of *Aedes albopictus* cell cultures persistently infected with Sindbis virus, *Virology*, 28, 69, 1977.
5. Igarashi, A., Characteristics of *Aedes albopictus* cells persistently infected with dengue viruses, *Nature*, 280, 690, 1979.
6. Schlesinger, R. W., Virus-host interactions in natural and experimental infections with alphaviruses and flaviviruses, in *The Togaviruses*, Schlesinger, R. W., Ed., Academic Press, New York, 1980, 83.
7. Bras-Herreng, F., Adaptation d'une population du virus Sindbis in *Drosophila melanogaster, Ann. Microbiol. (Inst. Pasteur)*, 127B, 541, 1976.
8. Igarashi, A., A mutant of chikungunya virus isolated from a line of Singh's *Aedes albopictus* cells by plaque formation on virus-sensitive cloned cells obtained from another Singh's *A. albopictus* cell line, *Virology*, 98, 385, 1979.
9. Hammon, W. McD., Rudnick, A., and Sather, G. E., Viruses associated with epidemic hemorrhagic fevers of the Philippines and Thailand, *Science,* 131, 1102, 1960.
10. Halstead, S. B., Mosquito-borne haemorrhagic fevers of South and Southeast Asia, *Bull. W.H.O.,* 35, 3, 1966.
11. Singh, K. R. P., Cell cultures derived from larvae of *Aedes albopictus* (Skuse) and *Aedes aegypti* (L.), *Curr. Sci.,* 36, 506, 1967.
12. Eagle, H., Amino acid metabolism in mammalian cell cultures, *Science,* 130, 432, 1959.
13. Mitsuhashi, J. and Maramorosch, K., Leafhopper tissue culture: embryonic, nymphal, and imaginal tissues from aseptic insects, *Contr. Boyce Thompson Inst.,* 22, 435, 1964.
14. Igarashi, A., Isolation of a Singh's *Aedes albopictus* cell clone sensitive to dengue and chikungunya viruses, *J. Gen. Virol.,* 40, 531, 1978.
15. Miles, J. A. R., Epidemiology of the arthropod-borne encephalitis, *Bull. W.H.O.,* 22, 339, 1960.
16. Igarashi, A., Buei, K., Ueba, N., Yoshida, M., Ito, S., Nakamura, H., Sasao, F., and Fukai, K., Isolation of viruses from female *Culex tritaeniorhynchus* in *Aedes albopictus* cell cultures, *Am. J. Trop. Med. Hyg.,* 30, 449, 1981.
17. Okuno, Y., Sasao, F., Fukunaga, T., and Fukai, K., An application of PAP (peroxidase-anti-peroxidase) staining technique for the rapid titration of dengue virus type 4 infectivity, *Biken J.,* 20, 29, 1977.
18. Voller, A., Bidwell, O., and Bartlet, A., Microplate enzyme immunoassay for the immunodiagnosis of viral infections, in *Manual of Clinical Immunology*, Rose, N. R. and Friedman, N., Eds., American Society of Microbiology, Washington, D.C., 1976. 506.
19. Bundo, K., Morita, K., and Igarashi, A., Detection of Japanese encephalitis virus by enzyme-linked immunosorbent assay, *Abstr. 31st Annu. Meeting of Japanese Virologists Society,* Osaka, 1983, 3019.
20. Sentsui, H. and Kono, Y., An epidemic of Getah virus infection among racehorses: isolation of the virus, *Res. Vet. Sci.,* 29, 157, 1980.
21. Igarashi, A., Makino, Y., Matsuo, S., Bundo, K., Matsuo, R., Higashi, F., Tamoto, H., and Kuwatsuka, M., Isolation of Japanese encephalitis and Getah viruses by *Aedes albopictus* clone C6/36 cells and by suckling mouse brain inoculation in Nagasaki, 1980, *Trop. Med.,* 23, 69, 1981.
22. Igarashi, A., Morita, K., Bundo, K., Matsuo, S., Hayashi, K., Matsuo, R., Harada, T., Tamoto, H., and Kuwatsuka, M., Isolation of Japanese encephalitis and Getah viruses from *Culex tritaeniorhynchus* and slaughtered swine blood using *Aedes albopictus* clone C6/36 cells in Nagasaki, 1981, *Trop. Med.,* 23, 177, 1981.
23. Igarashi, A., Sasao, F., Fukai, K., Buei, K., Ueba, N., and Yoshida, M., Mutants of Getah and Japanese encephalitis viruses isoalted from field-caught *Culex tritaeniorhynchus* using *Aedes albopictus* clone C6/36 cells, *Ann. Virol. (Inst. Pasteur)*, 132E, 235, 1981.
24. Igarashi, A., Detection of Mutant Viruses in Laboratory-Colonized Mosquitoes Infected with Arthropod-Borne Togaviruses, Rep. of the Grant in Aid for Scientific Research, Ministry of Education, Science, and Culture of Japan, 1983.
25. Igarashi, A., Sasao, F., Wada, T., Buei, K., Ueba, N., Yoshida, M., and Fukai, K., An enveloped virus isolated from field-caught *Culex tritaeniorhynchus* using *Aedes albopictus* clone C6/36 cell cultures, *Abstr. 5th Int. Congr. Virol.,* Strasbourg, 1981, 286.
26. Ueba, N., Minekawa, K., Yoshida, M., Kimura, A., Yuzaki, T., and Kitaura, T., Characteristics of enterovirus-like particles isolated from *Culex tritaeniorhynchus, Abstr. 31st Annu. Meeting of Japanese Virologist Society,* Osaka, 1983, 3010.
27. Okuno, Y., Igarashi, A., Fukunaga, T., Tadano, M., and Fukai, K., Electron microscopic observation of a newly isolated flavivirus-like virus from field-caught mosquitoes, *J. Gen. Virol.,* 65, 803, 1984.
28. Igarashi, A., Lin, W.-J., and Hayashi, K., An enveloped virus isolated from leafhoppers captured on East China Sea, *Trop. Med.,* 24, 179, 1982.

29. Henchal, E. A., Gentry, M. K., McCown, J. M., and Brandt, W. E., Dengue virus-specific and flavirirus group determinants identified with monoclonal antibodies by immunofluorescence, *Am. J. Trop. Med. Hyg.*, 31, 830, 1982.

30. Kuno, G., A continuous cell line of a nonhematophagous mosquito, *Toxorhynchites amboinensis, In Vitro*, 16, 915, 1980.

31. Shannon, J. E., Registry of Animal Cell Lines, 2nd ed., American Type Culture Collection, Maryland, 1962.

32. Igarashi, A., Fujita, N., Okuno, Y., Oda, T., Funahara, Y., Shirahata, A., Ikeuchi, H., Hotta, S., Wiharta, A. S., Sumarmo, and Sujudi, Isolation of dengue viruses from patients with dengue hemorrhagic fever (DHF) and those with fever of unknown origin (FUO) in Jakarta, Indonesia, in the years of 1981 and 1982, in *ICMR Annals, Kobe University*, Vol. 2, 1982, 7.

33. Tesh, R. B., A method for the isolation and identification of dengue viruses, using mosquito cell cultures, *Am. J. Trop. Med. Hyg.*, 58, 1053, 1979.

34. Rosen, L. and Gubler, D., The use of mosquitoes to detect and propagate dengue viruses, *Am. J. Trop. Med. Hyg.*, 23, 1153, 1974.

35. Fukunaga, T., Okuno, Y., Srisupaluck, S., Auvanich, W., Rojanasuphot, S., Sangkawibha, N., Kasemsarn, P., and Dharakul, C., Serological and virological studies on patients with dengue hemorrhagic fever (DHF) in Chanthaburi Province, Thailand. II. Serological characteristics of viruses isolated from DHF patients using a clone of Singh's *Aedes albopictus* cells, *Biken J.*, 23, 123, 1980.

36. Rojanasuphot, S., Chanyasanha, C., Ahandrik, S., Chatiyanonda, K., Igarashi, A., and Inouye, S., Isolation and identification of dengue viruses by combined use of C6/36 cells and the immune adherence hemagglutination test, *Jpn. J. Med. Sci. Biol.*, 34, 375, 1981.

37. Igarashi, A., Ogata, T., Fujita, N., Fukunaga, T., Mori, A., Uzuka, Y., Supawadee, J., Chiowanich, P., Peerakome, S., Leechanachai, P., Charoensook, O., and Chanyasanha, C., Flavivirus infections in Chiang Mai area, Thailand, in 1982, *Southeast Asian J. Trop. Med. Public Health*, 14, 470, 1983.

38. Zhang, Y.-h., Yu, W.-f., Tian, Z.-w., Ge, J.-q., Chen, Q.-s., and Wang, Y.-m., A simplified new method using enzyme immunoassay on cultured cells, *Microbiol. Immunol.*, in press.

39. Cunningham, A., Buckley, S. M., Casals, J., and Webb, S. R., Isolation of chikungunya virus contaminating an *Aedes albopictus* cell line, *J. Gen. Virol.*, 27, 97, 1975.

40. Hirumi, H., Viral microbial, and extrinsic cell contamination of insect cell cultures, in *Invertebrate Tissue Culture, Research Applications*, Maramorosch, K., Ed., Academic Press, New York, 1976, 233.

Chapter 9

ARTHROPOD CELL CULTURES IN STUDIES OF TICK-BORNE TOGAVIRUSES AND ORBIVIRUSES IN CENTRAL EUROPE

J. Reháček

TABLE OF CONTENTS

I. INTRODUCTION

The in vitro cultivation of cells and tissues from tick and mosquito vectors of arboviruses pathogenic for man and animals may aid the investigation of many problems relative to the relationships between these viruses and their arthropod hosts. Arthropod tissue cultures are being developed to determine their advantages, as compared to those of vertebrate cell cultures, with regard to greater amounts of available cells, higher sensitivity, and economy of the system. Virologists are using tick and mosquito cells in vitro for studies of Central European arboviruses, including the isolation of these viruses from nature and experimental studies on the relationship between these viruses and invertebrate cells. Results of these studies are reviewed here.

The presence of arboviruses in the families of Togaviridae and Reoviridae in Central Europe is determined either by direct isolation or by detection of antibodies in humans and animals. Of primary importance to humans is the togavirus, tick-borne encephalitis (TBE), of the arbovirus group B, genus *Flavivirus*, European subtype, and the reoviruses of the Kemerovo subgroup, genus *Orbivirus* (i.e., Kemerovo, Lipovnik, and Tribec viruses). These are tick-borne and probably widespread in Central Europe.

A. TBE Virus (Genus *Flavivirus*), European Subtype

TBE was the first arbovirus recovered in Central Europe, having been isolated in 1949 from *Ixodes ricinus* ticks in Czechoslovakia. This tick species is probably the main vector of TBE virus, as well as a long-term virus reservoir. Other natural reservoirs of TBE virus are various species of small vertebrates of which mammals (rodents and insectivores) are probably the most important. The virus is transmitted to man via tick bite and also by the consumption of raw goat and sheep milk, as well as the consumption of unpasteurized sheep milk products and, rarely, by aerosol. The TBE virus represents a serious public health problem in Central Europe.

B. Kemerovo, Tribec, and Lipovnik Viruses (Genus *Orbivirus*)

A member of the genus *Orbivirus* was first discovered during an expedition organized by the Institute of Poliomyelitis and Viral Encephalitides, Academy of Medical Sciences of the U.S.S.R., Moscow, and the Institute of Virology, Slovak Academy of Sciences, Bratislava, during 1962 in Western Siberia. The agent was first isolated from *Ixodes persulcatus* ticks and named the Kemerovo virus after the region where the field investigations had been done. In 1963 two similar viruses were isolated from *Ixodes ricinus* ticks collected in Slovakia. The latter had been found in two distant localities, Tribec and Lipovnik, and were named after those localities. The main vector of these viruses in Central Europe is the tick *Ixodes ricinus*, and the reservoirs are probably small terrestrial mammals. Man becomes infected following a tick bite and the disease is manifested as a benign meningo-encephalitis.

II. CULTIVATION OF TOGAVIRUSES IN ARTHROPOD CELLS IN VITRO

A. Cultivation of Tick-Borne Encephalitis (TBE) Virus in Tick Primary Cell Cultures
1. Cultures Prepared from *Hyalomma dromedarii* Ticks
a. Propagation of the Virus
With sufficient amounts of in vitro tick cells available[1] it is possible to use these cells for investigations with viruses. Thus, Rehacek[2-4] has shown that TBE virus multiplies well in tick primary cultures, including all cell types, without causing any cytopathic effect (CPE). The TBE viral strain used was the strain Hypr in its 43 to 48th intracerebral mouse passage. Tissue cultures from *Hyalomma dromedarii* ticks were seeded

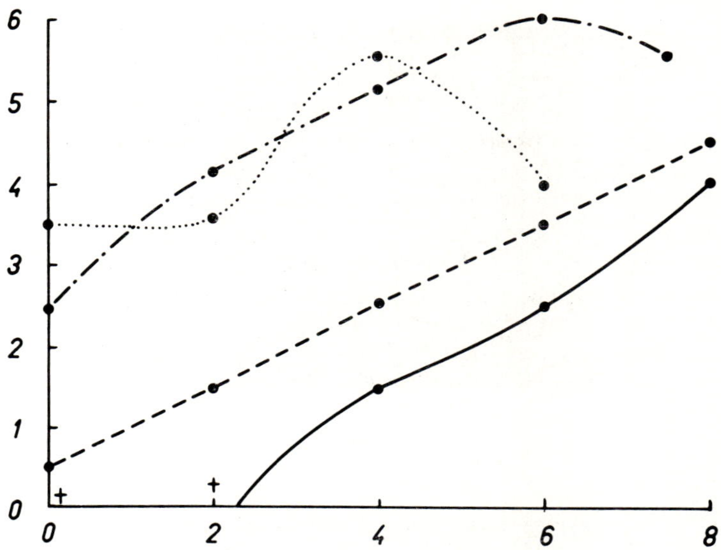

FIGURE 1.   Cultivation of mouse-passaged TBE virus (Hypr strain) in *Hyalomma dromedarii* tissue cultures. Abscissa, days of culture; ordinate, log $LD_{50}/0.03$ mℓ (mouse intracerebral route) + no virus found in medium.

into test tubes and, approximately 24 hr after seeding when cell growth was noted, the cultures were inoculated with three different doses of the virus. Samples of medium were removed from extracellular virus assay after 2, 4, 6, and 8 days and no fresh medium was added. Each sample was titrated separately by intracerebral inoculation of mice or by inoculation of chick embryo cells (CEC). Results are given in Figure 1. Each curve was computed from values obtained from two Carrel flasks. The highest titer of virus was obtained with a small inoculum (1 to 10 mouse $LD_{50}/0.03$ mℓ). The titer of virus increased by 1 to 5 log units, the virus increment being approximately of the order 0.5 to 1 log unit/day. In comparison with vertebrate-tissue cultures, the virus multiplied slowly in tick-tissue cultures, which may reflect the influence of a lower incubation temperature (27°C) on the multiplication of both the cells and the virus. It was also demonstrated that the TBE virus multiplied in cultures composed of hemocytes only, but the titers of virus thus obtained were not as high as in cultures containing all cell types.

The propagation of TBE virus in experimentally infected tick cells in vitro has also been studied by means of fluorescent antibody staining. Results have shown that virus propagation occurs only in the cytoplasm and is concentrated around the nucleus (Figures 2 and 3).[5] The positive fluorescence in tick cells in vitro and the virus titer in the cells is directly related to the time following the infection (Table 1).

b. Comparison of Susceptibility of Tick and Chick Embryo Cells (CEC)

It is generally known that CEC cultures appear to be the most susceptible of various cell systems to small doses of viruses of the TBE complex. We compared the sensitivity of this substrate with primary tick cultures for detection of small amounts of the TBE virus.[6]

Tube cultures of CEC were grown for 24 hr at 37°C in medium 199 supplemented with 10% unheated horse serum. Tube cultures of tick cells from *Hyalomma dromedarii* were prepared as previously described[2] and were also grown for 24 hr. Small amounts of the Hypr strain of the TBE virus in its 54th mouse passage were inoculated

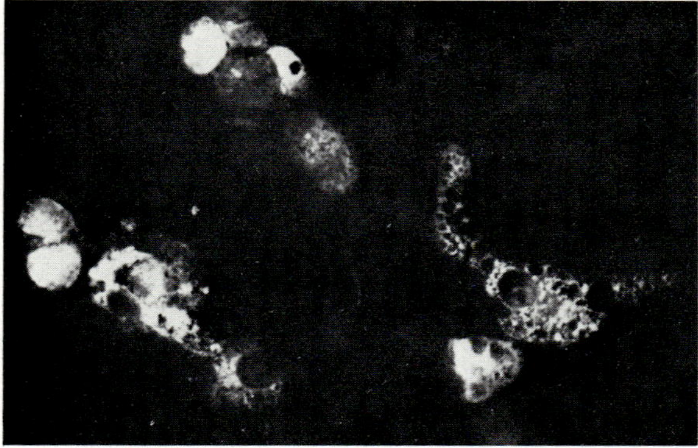

FIGURE 2

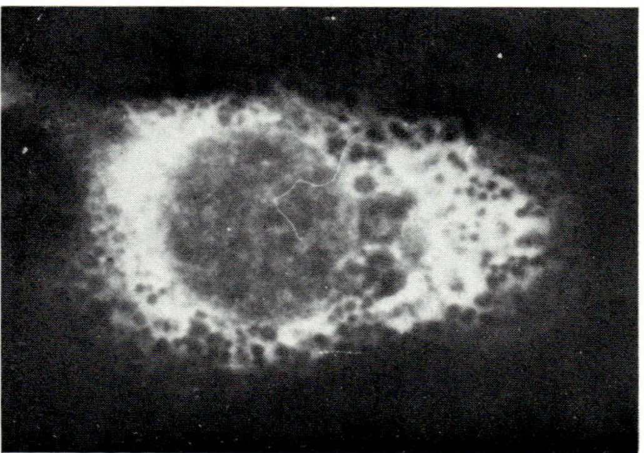

FIGURE 3

FIGURES 2 AND 3.   Multiplication of TBE virus in *H. dromedarii* cells in vitro revealed by immunofluorescence technique 8 days postinfection, magnification × 400.

into 4 to 10 tubes containing either CEC or tick cells. After 5 days, subcultures were made to CEC in which TBE virus was demonstrated by interference with WEE virus.

With inocula of 30 and 3 IFD_{50} the percentage of infected CEC and tick cell cultures was approximately equal (Table 2). With inocula containing 0.3 and 0.03 IFD_{50} of TBE virus, the respective percentages of infected tick cell cultures were 76 and 18 compared to that of CEC of 34 and 3. These results indicate that tick-cell cultures provide a more sensitive system than chick embryo cell cultures; the former is one of the most susceptible systems for detecting small amounts of TBE virus.

c. Isolation of the Virus from Miscellaneous Sources

Following this demonstration, we carried out experiments on the isolation of the TBE virus from blood samples, brain suspensions from small mammals, and suspensions made from ticks collected in natural foci of the virus in Czechoslovakia.[7]

Test materials were inoculated into 2- to 5-day-old tick-cell cultures without washing or changing of medium and incubated for 7 to 9 days. In CEC cultures inoculated with the same samples these were left to adsorb for 2 hr, after which the cultures were washed, fresh medium was added, and cultures were incubated for 5 days. At intervals

Table 1
DETECTION OF TICK-BORNE ENCEPHALITIS VIRUS IN PRIMARY TISSUE CULTURES PREPARED FROM *HYALOMMA DROMEDARII*

	Virus yields and immuno-fluorescence at days following infection		
Inoculum[a]	2	5	9
6.5	3.5[b]	5.5[b]	5.5[b]
2.5	1.5[b]	4.0[b]	6.5[b]
1.5	0.5	3.0[b]	4.5[b]
0.5	neg	2.0[b]	4.5[b]
0.05	neg	neg	5.0[b]
0.005	neg	neg	neg

[a] Log $LD_{50}/0.03$ mℓ (mouse intracerebral route).
[b] Positive immunofluorescence.

Table 2
COMPARISON OF THE SUSCEPTIBILITY OF TICK (*HYALOMMA ASIATICUM*) TISSUE AND CHICK EMBRYO CELL CULTURES IN DETECTING SMALL AMOUNTS OF TICK-BORNE ENCEPHALITIS VIRUS[a]

Inoculum[b] ($IFD_{50}/0.1$ mℓ)	Type of culture	Experiment[c]						Cultures infected (%)
		1	2	3	4	5	6	
30.00	TT	5/5	4/4	4/4	5/5	10/10	10/10	100
	CEC	5/5	4/4	4/4	5/5	10/10	10/10	100
3.00	TT	5/5	4/4	4/4	5/5	9/10	9/10	95
	CEC	5/5	2/4	4/4	5/5	9/10	10/10	92
0.30	TT	5/5	3/4	4/4	5/5	3/10	9/10	76
	CEC	2/5	1/4	2/4	5/5	1/10	2/10	34
0.03	TT	1/5	1/4	1/4	1/5	1/10	2/10	18
	CEC	0/5	0/4	0/4	1/5	0/10	0/10	3

[a] TT = tick tissue cultures; CEC = chick embryo cell cultures.
[b] IFD_{50} = a dose of tick-borne encephalitis virus interfering with WEE virus in half of the tubes inoculated.
[c] Numerator = number of tube-cultures infected; denominator = number of tube-cultures inoculated.

the culture fluids from both tick cell or CEC cultures were intracerebrally inoculated into suckling mice which were subsequently observed for 2 weeks. Brain tissue from mice which succumbed from suspected viral infection was used for further passages and the agents isolated were identified by neutralization tests carried out in 6 to 8 mice using intracerebral inoculation and anti-TBE hyperimmune goat serum (neutralization index = 10,000).

A total of 187 samples were examined from which 5 strains of TBE virus were isolated: 1 from the blood of *Talpa europaea*, 1 from the blood of *Apodemus flavicollis*,

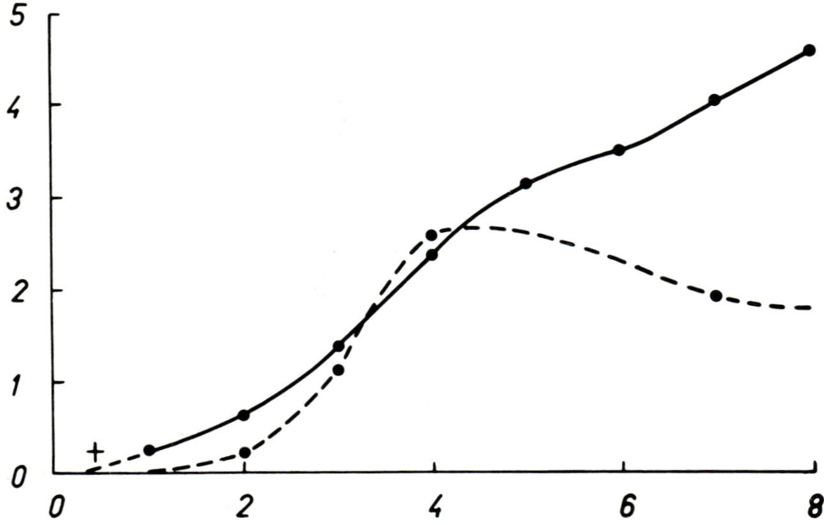

FIGURE 4.   Cultivation of HeLa cell-adapted TBE virus (Hypr strain) in tissue cultures of *Dermacentor marginatus*. Abscissa, days of culture; ordinate, $CPD_{50}/0.1$ mℓ (mouse intracerebral route) + no virus found in medium.

and 3 from *Ixodes ricinus*. All the strains were isolated by both of the methods. No toxic effects from either mammalian blood, brain suspensions, or tick suspensions were observed in either tick-cell cultures or CEC.

The results indicate that the method of isolating the TBE virus in tick-tissue cultures is equivalent to that observed in CEC cultures, and that tick-tissue cultures can be used for isolation of TBE virus in addition to suckling mice and CEC cultures in detection of virus from nature.

d. Maintenance of the Virus in Tick Cells

The TBE virus — Hypr strain — in its 46th mouse intracerebral passage has been maintained for long periods in tick cells in vitro.[8] Passages were made at weekly intervals by transferring 0.1 mℓ of supernatant fluid from previous cultures to new tubes containing growing tick cells. During a 1-year period 34 passages were performed. Each passage was titrated by intracerebral inoculation of mice and, at approximately every 5th passage, titrations were done by subcutaneous inoculation of mice. The virus persisted for 253 days in passaged cultures where its titers ranged between 3.0 and 6.5 log units during the passages. At the 6th, 10th, and 15th passages titers remained at the same level, and at the 20th passage the titer in subcutaneously injected mice was 0.5 log units higher. However, at the 25th and 30th passages these titers were 1 log unit lower. Although it is difficult to derive conclusions concerning any change in virus virulence, it can be concluded that it is possible to maintain the virus in tick cells as well as in other types of cells (i.e., vertebrate) in vitro.

2. Cultivation of the Virus in Cultures of Dermacentor marginatus

Primary cell cultures of *Dermacentor marginatus* ticks were used for the propagation of the TBE virus, Hypr strain adapted to HeLa cells.[3] This virus, despite its adaptation to HeLa cells, multiplied well in *Dermacentor marginatus* cells (Figure 4). In a few experiments the cells did not adhere to the glass but survived as free cells suspended in the medium. Although the virus multiplied in these cells, its increase was only 1 log unit as compared with cell cultures in the same experiment in which the amount of the

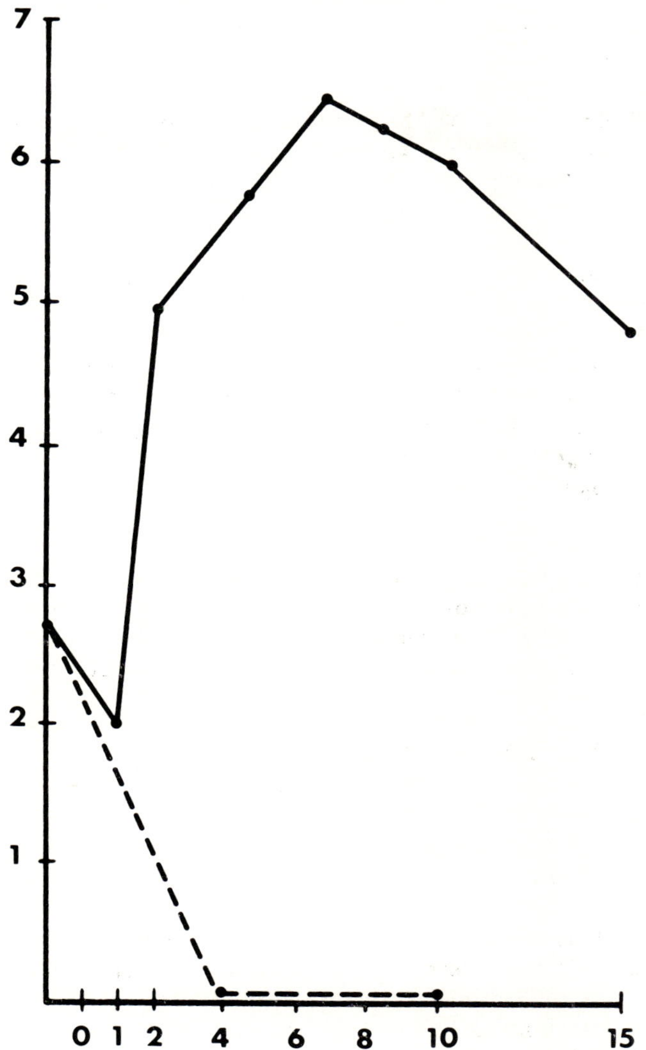

FIGURE 5.   Multiplication of TBE virus (Hypr strain) in a *D. paru-mapertus* cell line. Abscissa, days of culture; ordinate, virus titer (PFU/mℓ in Vero cells).

virus in stationary and growing cells increased by 5 log units. This indicates that only growing cells can support marked multiplication of the TBE virus. Virus propagation in the tick cells occurred without any CPE. In several experiments the dosages of virus used for the inoculation of the tick-tissue cultures were so small that they were not detectable in intracerebrally inoculated mice or in HeLa cells. This observation could possibly support the use of tick cells for virus isolation experiments.

B. Cultivation of TBE Virus in Tick Cell Lines
1. Dermacentor parumapertus Cells

A line of *Dermacentor parumapertus* (RML-14) cells[9] from passages 10 to 17 was successfully used for the cultivation of the TBE virus, Hypr strain, the 2nd mouse passage, plus an unstated number of passages before experiments with tick cells were done.[10] The virus multiplied at a relatively high level and a maximum titer of 4.8 to 6.6 dex was attained between the 4th and 10th day after the infection (Figure 5).

C. Cultivation of TBE Virus in Mosquito Cell Lines
1. Anopheles gambiae Cells

The cell line Mos. 55, derived from the larvae of *Anopheles gambiae* mosquitoes by Marhoul and Pudney[11] was used for the propagation of the TBE virus (Hypr strain, 39th mouse passage). The 50 to 54th passage of cells was used in these experiments. The virus did not adsorb on the cells and did not multiply; only traces of virus were detectable until the 5th day postinfection (p.i.) in the system used.[12]

2. Aedes albopictus Cells

The *Aedes albopictus* cell line of Singh,[13] in its 41st and 48th subcultures, was used for the cultivation of the TBE virus. Strain "40" of the TBE virus which was isolated from *Ixodes ricinus* ticks and in its 3rd intracerebral mouse passage was tested. The virus was detected in culture fluids at 3 days p.i. only at a very low level (less than 0.5 log LD_{50}) and evidently did not multiply.[14] In another study the Hypr strain of the TBE virus (passage level unknown) failed to plaque in Singh's *Aedes albopictus* cells in their 87th to 135th subculture.[15]

3. Aedes aegypti Cells

The *Aedes aegypti* cell line of Varma and Pudney[16] Mos. 29, in the 76th passage was employed for the cultivation of the TBE virus, strain "40", which was in its 3rd intracerebral passage in mice.[17] The virus did not multiply in mosquito cells and showed only a gradual decrease in titer from the 2nd day p.i., disappearing completely within 14 days p.i.

III. CULTIVATION OF ORBIVIRUSES IN ARTHROPOD CELLS IN VITRO

A. Cultivation of Viruses in Tick Primary Cell Cultures
1. Cultivation of Tribec Virus in Primary Cultures of Hyalomma dromedarii

The Tribec virus was successfully cultivated in *Hyalomma dromedarii* tissue cultures by Rehacek et al.[18] Tissue cultures from *Hyalomma dromedarii* ticks were prepared as previously described.[1] Cells were seeded on coverslips in test tubes and the monolayers that formed in 3 to 5 days were infected with virus without previous washing. Samples of $^1/_{10}$ mℓ of the medium were taken from the infected tissue cultures at various intervals and the virus titers determined by inoculation of suckling mice. The virus in the medium dropped to undetectable values 24 hr after inoculation. Starting from the 2nd day p.i., the titer increased about 4 log units and remained at this level until the 14th day when the experiment was terminated (Figure 6).

In parallel with infectivity titrations, the occurrence of a viral antigen in the infected tick cells was traced by fluorescent antibody methods. This procedure employed the indirect method with immune mouse serum prepared against Tribec virus and antimouse conjugate. A clear-cut specific fluorescence was seen in the cytoplasm from the 2nd to 7th day p.i. At the beginning, only some fluorescing granules with perinuclear localization were found, but by the 4th day they increased in number and blended together into large bright fluorescing masses filling the whole cytoplasm (Figures 7 to 10). A viral antigen was detected in both epithelial- and fibroblast-like cells. The amount of extracellular infective virus was qualitatively related to the intensity of the cytoplasmic fluorescence and it can be assumed that the amount of intracellular infective virus surpassed the critical value of 10^4 mouse LD_{50} per 0.01 mℓ cell suspension which is necessary for the immunofluorescent detection of Tribec virus in CEC.

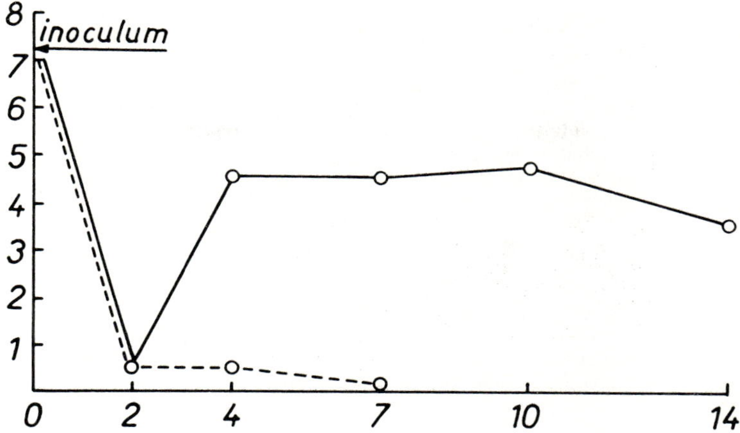

FIGURE 6.   Multiplication of Tribec virus in *H. dromedarii* tissue cultures. Abscissa, days of culture; ordinate, log $LD_{50}/0.1$ mℓ (mouse intracerebral route).

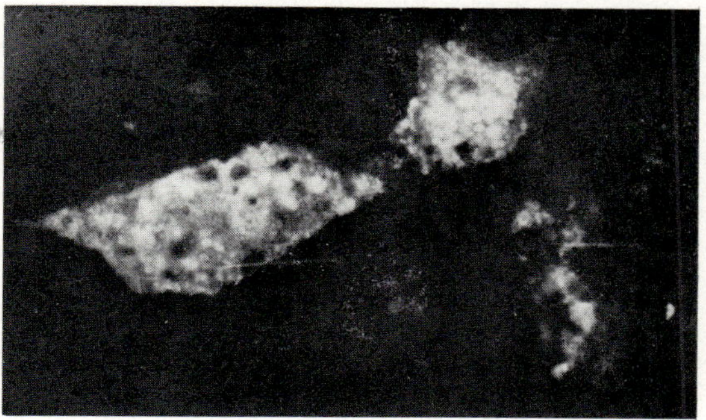

FIGURE 7

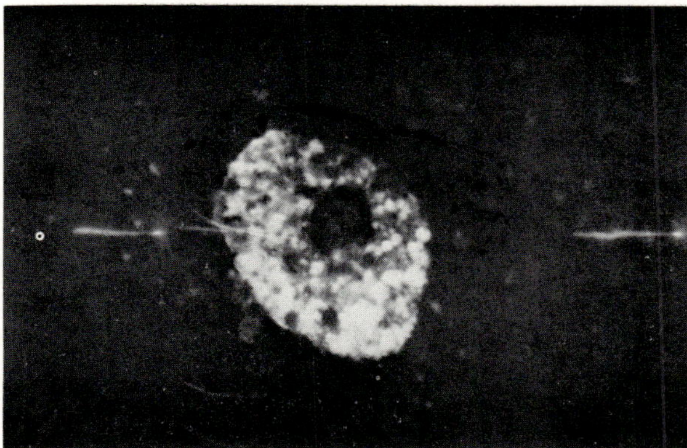

FIGURE 8

FIGURES 7 THROUGH 10.   Multiplication of Tribec virus in *H. dromedarii* tissue cultures 4 days postinoculation, magnification × 400.

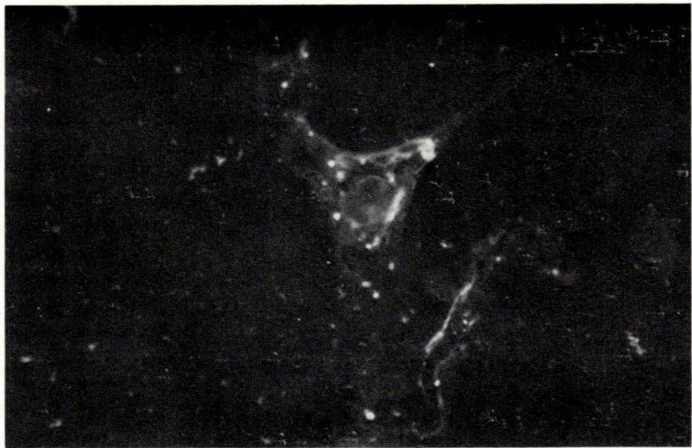

FIGURE 9

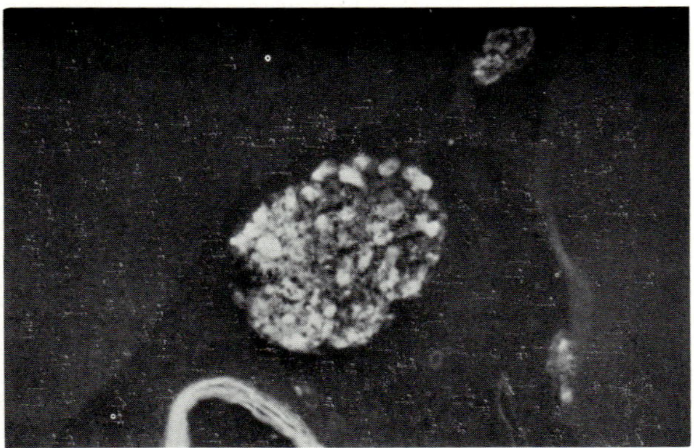

FIGURE 10

B. Cultivation of Viruses in Tick Cell Lines
1. Kemerovo Virus in Dermacentor variabilis Cells

The cells of two *Dermacentor variabilis* lines, RML-15 and RML-19 were shown to be susceptible to infection with the Kemerovo virus, R10 strain. The virus grew in these cells at 37°C without CPE. Of the RML-15 line, titer of the virus (as assayed in Vero cells) was $10^{4.0}$ PFU/mℓ in Vero cells[19] 1 week after infection.

2. Kemerovo and Tribec Viruses in Dermacentor parumapertus Cells

Cultivation of Kemerovo virus (R10 strain) at the 8th mouse passage and Tribec virus (VR468 strain) at the 18th mouse passage was carried out in the *Dermacentor parumapertus* (RML-14) cell line.[10] These cells were in their 10th to 17th passages. Maximum titers of $10^{5.1}$ for the Kemerovo virus and $10^{5.3}$ for the Tribec virus were obtained on days 10 and 4, respectively (Figures 11 and 12).

C. Cultivation of Viruses in Mosquito Cell Lines
1. Propagation of Viruses

Buckley[20] first cultivated Kemerovo 10, Lipovnik 91, and Tribec (original) viruses in Singh's *Aedes* cell lines which were maintained at 35°C. Kemerovo virus infected both *A. aegypti* and *A. albopictus* cells similarly, but Lipovnik and Tribec viruses showed higher titers in the *A. albopictus* cell line than in the *A. aegypti* cell line (Table 3).

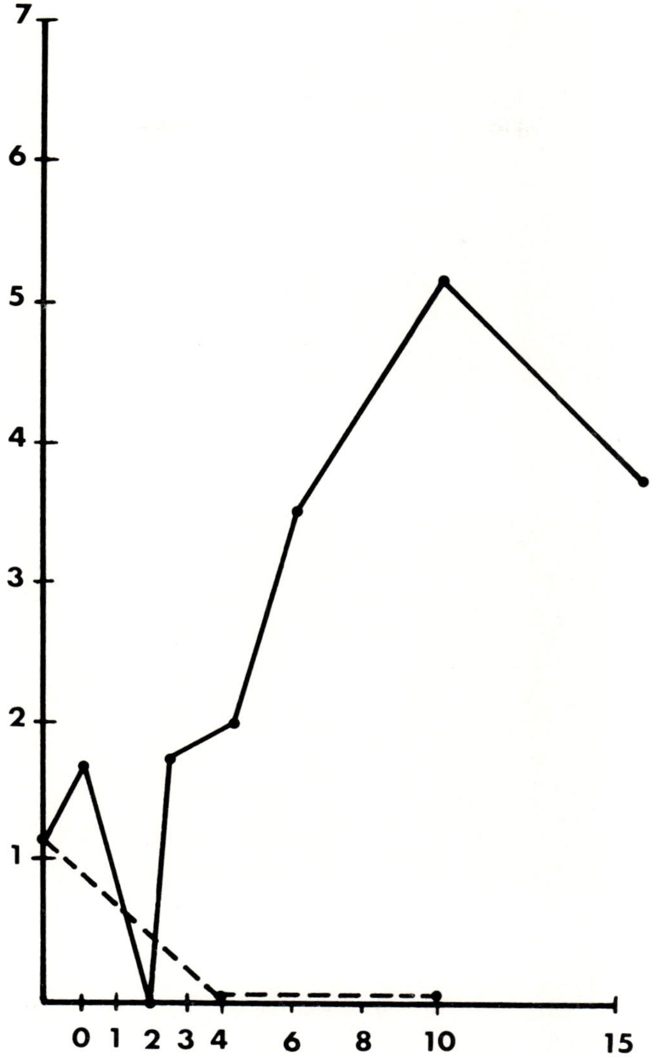

FIGURE 11.   Multiplication of Kemerovo virus (strain R10) in a *D. parumapertus* cell line. Abscissa, days after inoculation; ordinate, virus titer (PFU/m*l* in Vero cells).

In further experiments *A. albopictus* and *A. aegypti* cells were used in their 163rd to 205th and 57th passage, respectively, for the cultivation of Kemerovo virus strain L75, which had been isolated from a patient with meningitis.[21] This strain had undergone seven yolk sac passages in chick embryos and three intracerebral passages in baby mice.

The virus multiplied in both cell lines without causing CPE. However, the results demonstrated a 2 to 4 dex lower sensitivity of insect cells as compared to BHK-21 and Vero cells. The *A. albopictus* cells incubated at 30°C had shown the virus to be present in the medium 7 days p.i. without significant increase in titer. At 35°C an increase of the virus titer from 1 to 2 dex was observed between the 2nd and 11th day p.i. Figure 13 shows the growth curve of the Kemerovo virus in stationary tube cultures of *A. albopictus* cells inoculated with a multiplicity of infection of 10 CPD_{50} of virus per cell. The yield of virus in these cells was very low, consisting of about 1 PFU/50 cells. Virus

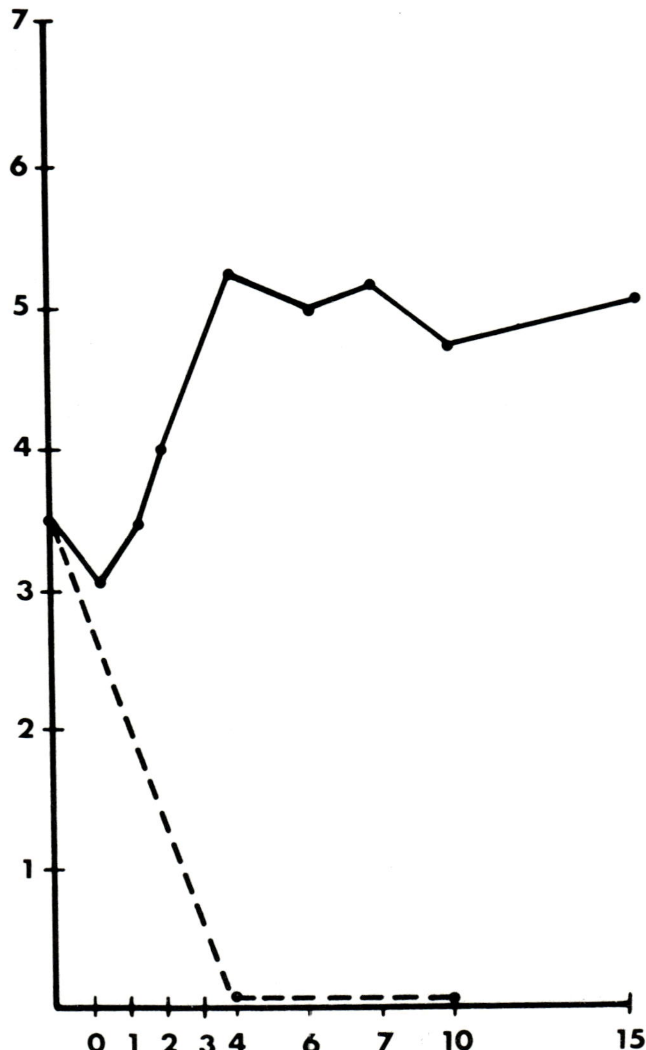

FIGURE 12.   Multiplication of Tribec virus (strain VR468) in a *D. parumapertus* cell line. Abscissa, days after inoculation; ordinate, virus titer (PFU/m*l* in Vero cells).

production was not improved when tubes were inoculated with virus and simultaneously seeded with cells.

2. Persistent Infection of Aedes albopictus Cells with Kemerovo Virus

Primary infection of these cells demonstrated a low virus yield indicating that only a small number of cells became infected.[21] Intervals between single passages varied between 3 to 8 days and a total of over 50 transfers in a period of 18 months were made. Transfer titers showed poor replication of virus ranging from 2 to 3 dex (Figure 14). Only a small number of cells (1.4 to 3.8%) were found infected at the 35th transfer when tested by plaque assay in Vero cells and by fluorescent antibody techniques. Immunofluorescence was visualized in the form of single or multiple granules or as diffuse specific cytoplasm staining.

Table 3
COMPARISON OF TITERS OBTAINED WITH KEMEROVO SUBGROUP VIRUSES IN SINGH'S LINES OF *AEDES AEGYPTI* AND *AEDES ALBOPICTUS* CELLS

Virus	Cell line	ID_{50}/ml (neg log_{10})	Day of subinoculation
Kemerovo	*Aedes aegypti*	6.3	6
	A. aegypti	5.5	14
	A. albopictus	5.7	6
	A. albopictus	6.2	14
Lipovnik	*A. aegypti*	4.3	4
	A. aegypti	3.7	7
	A. albopictus	4.7	4
	A. albopictus	5.3	7
Tribec	*A. aegypti*	5.7	6
	A. aegypti	4.5	14
	A. albopictus	7.5	6
	A. albopictus	7.5	14

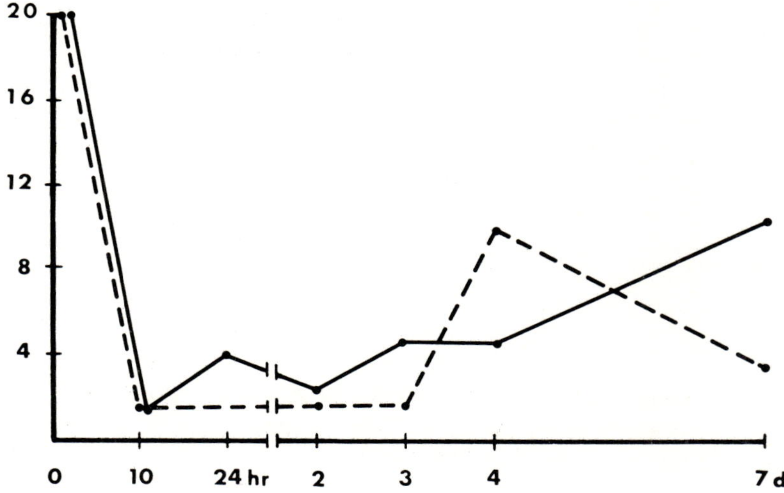

FIGURE 13.   Multiplication of Kemerovo virus (strain L75) in *Aedes albopictus* cells in vitro. Abscissa, time interval postinoculation; ordinate, virus titer (PFU/m*l* in Vero cells × 10³ PFU per tube). ————— is intracellular virus, ——— is extracellular virus.

No difference was noted between carrier culture virus at the 21st transfer, 5 months p.i., and parent strain in the ability to induce CPE in BHK-21 cells and plaques in Vero cells, although both CPD_{50} and PFU titers of the carrier cultures were lower. While the virus in carrier cultures showed a slight decrease in pathogenicity for baby mice, viral invasiveness as indicated by differences between intracerebral and subcutaneous titers showed no change.

3. Attempts to Detect Interferon in Aedes albopictus Cells

Carrier cultures of *Aedes albopictus* cells infected with Kemerovo virus (7th transfer) were successfully superinfected with Chikungunya virus.[21] Although some preparations caused a slight decrease of Chikungunya virus yields (from 50 to 1%) as compared to

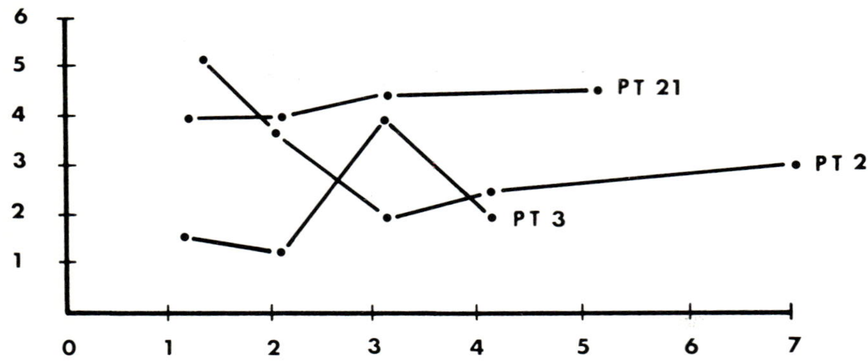

FIGURE 14.   Titers of extracellular (PT2 and PT3) and total virus (PT21) during individual transfers of *A. albopictus* cells persistently infected with Kemerovo virus (strain L75). Abscissa, days after transfer; ordinate, log PFU/m*l* (Vero cells).

controls, this effect was not regularly reproducible and no definite evidence of interferon-like substances was seen.

4. Attempts to Plaque Viruses in Aedes albopictus Cells

Tribec virus (VR468 strain, 18th mouse passage) formed plaques in *Aedes albopictus* cells, but Kemerovo virus (R10 strain, 7th mouse passage) and Lipovnik virus (strain 91, 1st CEC plus 3rd mouse passages) did not.[15] Tribec virus yielded the best plaques when the cultures were incubated at 35°C and read on postinoculation day 6. The average plaques measured 2.0 mm and the titer ($\log_{10}$PFU/m*l*) in the cells was 8.2. It is of interest to note that of 53 tick-borne agents tested, Tribec virus was the only one to produce plaques in the mosquito cells.

IV. MIXED INFECTION WITH TBE AND KEMEROVO VIRUSES IN TICK-TISSUE CULTURES

The problem of dual infections with arboviruses in the same biological vector became evident with the discovery of mixed foci of arbovirus infections. In eastern Slovakia, TBE virus was found in *Ixodes ricinus* ticks together with viruses of the Kemerovo subgroup.[22] This finding stimulated experiments to determine if mixed infections of both viruses could occur in tick cells in vitro. As a model, cell cultures prepared from developing adults of *Hyalomma dromedarii* ticks[1] were infected with the Hypr strain of TBE virus (50th mouse passage) and 2 strains of Lipovnik virus (strain 91 and strain 3) each in multiple CEC, chick embryo, and rat brain passages. TBE virus was titrated in mice (intracerebral route) and the Lipovnik viruses were titrated by plaque assay or CPE in CEC. Titration of viruses in dually infected cultures was accomplished by using immune serum prepared against the second virus.

In the first two experiments tick cells cultivated in plastic flasks were exposed to 10^3 LD_{50} of TBE virus for 2 hr and after 7 to 15 days of cultivation at 30°C they were superinfected with 10^4 PFU of Lipovnik virus. In the first experiment Lipovnik virus was added to the cultures without changing the medium; in the second the medium was removed before superinfection and a change of medium was made 2 hr after absorption of the superinfecting virus. Thereafter, 1.0 m*l* of medium was withdrawn at 3-day intervals and replaced with an equal amount of fresh medium. Titrations of the extracellular virus showed that no interference occurred between the TBE and Lipovnik viruses in the cultures tested (Figures 15 and 16). The percentage of cells infected with

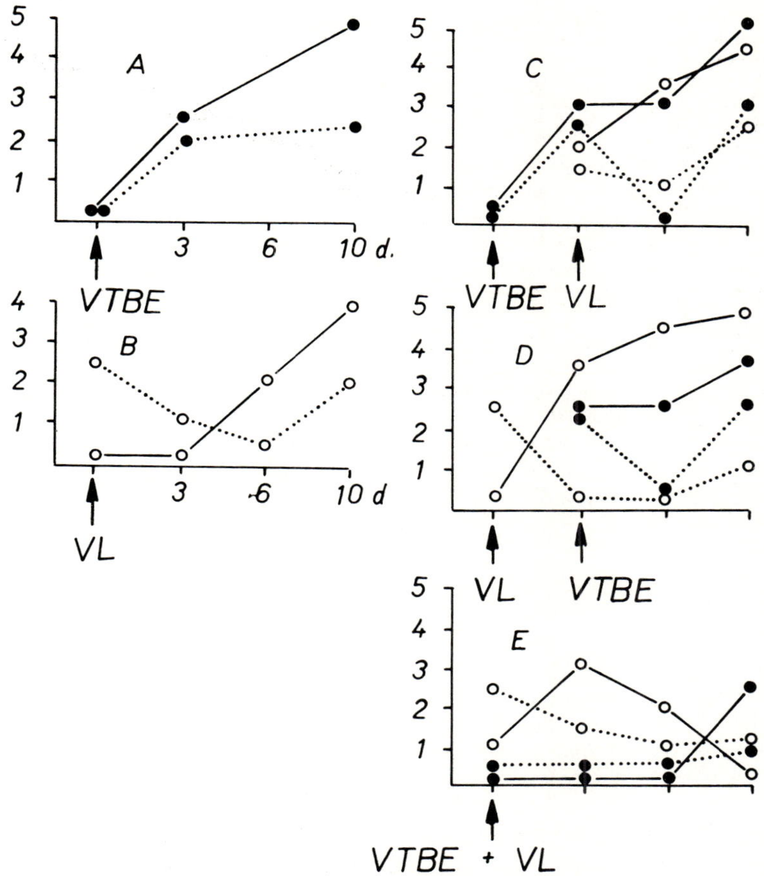

FIGURE 17. Various modifications of mixed infection with TBE and Lipovnik viruses in *H. dromedarii* tissue cultures. A is cultures infected with VTBE; B is cultures infected with VL; C, D, and E are dual infection with VTBE and VL. ● VTBE, O VL, ——— intracellular virus, and ······ extracellular virus (refer to Figure 15 caption for explanation).

REFERENCES

1. Reháček, J., Preparation of tissue cultures from the tick *Hyalomma dromedarii* Koch, *J. Med. Entomol.*, 2, 161, 1965.
2. Reháček, J., An improved method of tick tissue cultures, *Acta Virol.*, 6, 188, 1962.
3. Reháček, J., Propagation of tick-borne encephalitis virus (TBE) in tick tissue cultures, *Ann. Epiphyt.*, 14, 199, 1963.
4. Reháček, J., Cultivation of different viruses in tick tissue cultures, *Acta Virol.*, 9, 332, 1965.
5. Reháček, J., Tissue cultures of blood-sucking arthropods and their use for the cultivation of viruses and rickettsiae, *Ann. Parasitol. Hum. Comp.*, 46, 197, 1971.
6. Reháček, J. and Kozuch, O., Comparison on the susceptibility of primary tick and chick embryo cell cultures to small amounts of tick-borne encephalitis virus, *Acta Virol.*, 8, 470, 1964.
7. Reháček, J. and Kozuch, O., The use of *Hyalomma dromedarii* tick tissue cultures for the isolation of tick-borne encephalitis virus from its natural foci, *Acta Virol.*, 13, 253, 1969.
8. Reháček, J., Maintaining of tick-borne encephalitis (TBE) virus, western subtype, in tick cells *in vitro*, in *Proceedings Third International Colloquium on Invertebrate Tissue Culture*, Rehacek, J., Blaskovic, D., and Hink, W. F., Eds., Slovak Academy of Sciences, Bratislava, 1973, 439.

9. Bhat, U. K. M. and Yunker, C. E., Establishment and characterization of a diploid cell line from the tick, *Dermacentor parumapertus* Neumann (Acarina:Ixodidae), *J. Parasitol.*, 63, 1092, 1977.
10. Bhat, U. K. M. and Yunker, C. E., Susceptibility of a tick cell line *(Dermacentor parumapertus* Neumann) to infection with arboviruses, in *Arctic and Tropical Arboviruses,* Kurstak, E., Ed., Academic Press, New York, 1979, 263.
11. Marhoul, Z. and Pudney, M., A mosquito cell line (Mos. 55) from *Anopheles gambiae* larvae, *Trans. R. Soc. Trop. Med. Hyg.,* 63, 102, 1972.
12. Marhoul, Z., Susceptibility of *Anopheles gambiae* mosquito cell line (Mos. 55) to some arboviruses, *Acta Virol.*, 17, 507, 1973.
13. Singh, K. R. P., Cell cultures derived from larvae of *Aedes albopictus* (Skuse) and *Aedes aegypti* (L.), *Curr. Sci.,* 36, 506, 1967.
14. Danielova, V., Susceptibility of *Aedes albopictus* mosquito cell line to some arboviruses, *Acta Virol.,* 17, 249, 1973.
15. Yunker, C. E. and Cory, J., Plaque production by arboviruses in Singh's *Aedes albopictus* cells, *Appl. Microbiol.,* 29, 81, 1975.
16. Varma, M. G. R. and Pudney, M., The growth and serial passage of cell lines from *Aedes aegypti* L. larvae in different media, *J. Med. Entomol.,* 6, 432, 1969.
17. Malkova, D. and Marhoul, Z., Susceptibility of mosquito cell line *Aedes aegypti* to some arboviruses, in *Proceedings Third International Colloquium on Invertebrate Tissue Culture,* Rehacek, J., Blaskovic, D., and Hink, W. F., Eds., Slovak Academy of Sciences, Bratislava, 1973, 267.
18. Reháček, J., Rajcani, J., and Gresikova, M., Relationship of Tribec virus to tick cells and tissues, *Acta Virol.,* 13, 439, 1969.
19. Yunker, C. E., Cory, J., and Meibos, H., Continuous cell lines from embryonic tissues of ticks (Acari:Ixodidae), *In Vitro,* 17, 139, 1981.
20. Buckley, S. M., *In vitro* propagation of tick-borne viruses other than group B arboviruses, in *Proc. Int. Symp. Tick-borne Arboviruses,* (excluding group B), Gresikova, M., Ed., Veda Publishing House of the Slovak Academy of Sciences, Bratislava, 1971, 43.
21. Lubikova, H. and Buckley, S. M., Studies with Kemerovo virus in Singh's *Aedes* cell lines, *Acta Virol.,* 15, 393, 1971.
22. Libikova, H., Reháček, J., and Somogyiova, J., Viruses related to the Kemerovo virus in *Ixodes ricinus* ticks in Czechoslovakia, *Acta Virol.,* 9, 76, 1965.
23. Libikova, H., Reháček, J., and Rajcani, J., Attempts to prove interference between tick-borne encephalitis and Lipovnik viruses in tick tissue cultures, *Cesk. Epidemiol. Mikrobiol. Immunol.,* 23, 332, 1974.
24. Libikova, H. and Reháček, J., Attempts to prove interference between arboviruses in arthropod tissue culture, in *Interferon and Interferon Inducers,* Foldes, I. and Talas, M., Eds., Kiadta a MTA Mikrobiologie, Kutatocsoport, 1976, 177.
25. Reháček, J. and Libikova, H., Mixed infection with two tick-borne viruses in tick tissue cultures and attempts to demonstrate interference, in *Proceedings 4th International Congress of Acarology,* Piffl, E., Ed., Akademiai Kiado, Budapest, 1979, 401.

Chapter 10

NAIROBI SHEEP DISEASE VIRUS AND *REOVIRUS*-LIKE PARTICLES IN THE TICK CELL LINE TTC-243 FROM *RHIPICEPHALUS APPENDICULATUS:* EXPERIENCES WITH THE HANDLING OF THE TICK CELLS, IMMUNOPEROXIDASE, AND ULTRAHISTOLOGICAL STUDIES

E. Munz, M. Reimann, and H. Mahnel

TABLE OF CONTENTS

I. THE MULTIPLICATION OF BUNYAVIRUSES (FAMILY: BUNYAVIRIDAE) IN TICK CELL LINES

The multiplication of arboviruses in cultures of arthropod cells has evoked great interest and stimulated many aspects of arbovirus research. The use of different mosquito cell cultures has brought significant progress in fundamental and applied arbovirus research, fulfilling expectations at least partially. However, the in vitro cultivation of tick cells has not yet found such broad application. Although basic requirements for tick cell cultivation have been known for more than 10 years, research on these cells and their application to virus multiplication is carried out by only a few laboratories. Evidently, the in vitro cultivation of tick cells is more difficult than that of mosquito cells. Despite recently published valuable technical hints for the preparation and handling of tick cells,[1,2] uncertainties still exist and successful initiation of tick cell cultures, as well as establishment of continuous tick cell lines, are not always guaranteed. This may be one of the reasons why only a relatively small number of publications concerned with the multiplication of viruses in tick cell lines has appeared.

In 1975, Varma et al.[3] established the first serially propagating, continuous cell lines from ticks, *Rhipicephalus appendiculatus.* Following this, additional cell lines were established from ticks of 4 other genera, *Boophilus, Dermacentor, Haemaphysalis,* and *Rhipicephalus* (see Volume I, Chapter 3). Soon after the establishment of the first tick cell line from *R. appendiculatus,* Pudney et al.[4] reported the successful propagation of bunyaviruses in these cell lines. Growth was obtained with Zirqa (Hughes group), Keterah, and Lanjan (Kaisodi group) viruses. They then confirmed the growth of Ketapang, Dugbe, and Ganjam viruses to varying degrees.[5] Germiston, Tahyna, and California encephalitis, 3 mosquito-borne bunyaviruses, failed to replicate.

David-West[6] demonstrated multiplication of the Dugbe virus in the *Boophilus (B.) microplus* cells of Pudney et al.[7] These studies were extended by Leake et al.,[8] who investigated the susceptibility for several arboviruses of the tick cell line TTC-243 from *Rhipicephalus appendiculatus* and one from *B. microplus.* The authors confirmed the growth of Bunyamwera, Dugbe, and Ganjam viruses, but failed to detect multiplication of Germiston, California encephalitis, Tahyna, and Anopheles A viruses. The following viruses (at that time known as unclassified arboviruses) have shown multiplication: Keterah, Zirqa, Punta Salinas, Soldado, and Hughes. Now, the Hughes and Soldado viruses are regarded as bunyaviruses. The growth kinetics of all of these viruses varied considerably. These authors found that the Punta Salinas virus grows comparatively slowly. Titers reached 5.2 log 10 PFU/mℓ. The *B. microplus* cell line supported the growth of Zirqa, but not of the Bunyamwera virus. No cytopathic effect (CPE) has been seen in these cells with any of these viruses, and this is regarded as restricting the use of tick cells for primary isolation of viruses from field materials. Leake et al.[8] also reported the persistent infections of tick cells with nonbunyaviruses, but latent and extrinsic virus contamination in tick cells could not be found either in suckling mice, mammalian tissue cultures, or in an amphibian cell line by electron microscopy.

Bhat and Yunker[9] tested the viral susceptibility of the tick cell line RML-14 (from *Dermacentor (D.) parumapertus):* Hazara (Crimean-Congo hemorrhagic fever serogroup) and Dugbe (Nairobi sheep disease serogroup) viruses multiplied well, and maximum titers of 5.5 and 5.4 dex were attained on days 15 and 6 postinoculation, respectively. Hazara virus, which was inoculated at a concentration too low to detect in Vero cells, remained undetected in *D. parumapertus* cells up to day 4 after inoculation. Thereafter, its titer increased rapidly up to day 15. Silverwater (Kaisodi serogroup), Bhanja, and Sunday Canyon viruses persisted in *D. parumapertus* cells, but their growth, if any, remained undetected due to high doses of inocula that were used.

Hughes (Hughes serogroup) and Midway (Nyamanini serogroup) viruses multiplied well, whereas the Soldado (Hughes serogroup) virus showed only a low level of multiplication. In contrast, the Sapphire II (Hughes serogroup) virus showed only a gradual decrease in titer following inoculation. The authors found that all viruses associated with ixodid ticks multiplied well or were maintained at high levels in these cells, but only 1 argasid virus grew well. It was confirmed that tick cells permit growth of both mosquito- and tick-borne viruses irrespective of essential lipids, which is in contrast to the most extensively tested mosquito cell line, *Aedes albopictus* (ATC-15).

Response of the latter to viral infection depends primarily upon the vector of the virus, and is also correlated with the possession or lack by the virion of essential lipids.[10] Thus, ATC-15 cells readily support growth of mosquito-borne viruses regardless of their sensitivity to lipid solvents, but are refractory to tick-borne viruses, except orbiviruses, which are relatively resistant to lipid solvents. All findings to date indicate that the type of vector and presence or absence of essential lipids are apparently not eminent factors in determining infectivity of viruses for tick cells. The broad susceptibility of the tick cell line RML-14 of Bhat and Yunker[9] to vector-borne animal viruses does not extend to those nominally classified as arboviruses but which apparently do not utilize an intermediate vector. It is therefore believed that tick cell lines, like those from mosquitoes, may be useful in the identification, classification, and characterization of viruses.

According to Hink,[11] Ganjam, Bhanja, and Kaisodi viruses multiply in the *Haemaphysalis spinigera* cell line ATC-309. An extensive review of replication of arboviruses in arthropod in vitro systems was published by Pudney et al.,[12] where 42 bunyaviruses (including those classified as *Bunyavirus*-like) were cited. Of five bunyaviruses isolated from mosquitoes, only one, Bunyamwera, multiplied in tick cells and then only to a very low level. The tick-borne bunyaviruses did not grow in mosquito cells, apart from Ganjam, which has also been isolated from mosquitoes, but nine of thirteen multiplied in tick cells with two more being maintained at a low level. The mosquito-borne bunyaviruses did not grow in tick cells as readily as mosquito-borne togaviruses. This review clearly indicates that comparative investigations of arbovirus behavior in tick cells are feasible, but the authors point out that different results might have been obtained with virus material of heterogeneous origin.

Munz et al.[13] studied the susceptibility of the tick cell line TTC-243 for Nairobi Sheep Disease Virus (NSDV), genus *Nairovirus,* family Bunyaviridae, as well as other arboviruses and nonarboviruses. Nonarboviruses failed to multiply in the TTC cells incubated at 28°C, while the Sindbis virus (family Togaviridae) yielded evidence of multiplication without CPE.

The tick *Rhipicephalus appendiculatus* acts in nature as the main vector of NSDV, which is capable of causing severe epidemics and significant morbidity of sheep and goats. Nairobi sheep disease (NSD) is, consequently, of great importance in some countries of East Africa. Clinical signs include fever and severe gastroenteritis resulting in bloody diarrhea. Pregnant ewes may abort. Up to 50% mortality may be observed in new sheep moved into enzootic areas. Native sheep apparently are more susceptible than Merinos. The disease may also be fatal for goats (up to 10% mortality), although subclinical infections are more commonly observed in goats.

Tick cell cultures inoculated with NSDV yielded an abundant virus not only during the primary infection, but also during 35 passages, with titers ranging from 10^5 to 10^6 LD$_{50}$/mℓ. There was no CPE discernible at any time. Changes such as cytoplasmic inclusion bodies were not observed in conventionally stained monolayers grown on coverslip cultures.

Sheep infected with NSDV propagated in TTC-243 cells showed a typical febrile reaction on day 2 or 3 following inoculation. Gastroenteritis with diarrhea was not observed or was very mild. One African fat-tailed sheep and one Merino sheep inoculated with the second and fourth passage of NSDV in TTC-243 cells, respectively, developed viremia with a serum infectivity titer of $10^{2.7}$ $LD_{50}/m\ell$, as determined in infant mice. Another sheep inoculated with the 26th passage material failed to develop viremia. Inasmuch as the titers of the second, fourth, and twenty-sixth passages of NSDV in tick cells were of similar magnitude, loss of virulence for mice apparently did not occur. However, some degree of viral attenuation for sheep seems to be a distinct possibility. The main objectives of these virus-susceptibility studies were twofold. Generally, we were interested in an in vitro biological characterization of arboviruses and nonarboviruses in tick cells. Specifically, our goal was to obtain a less virulent, yet still immunogenic NSDV strain for sheep and goats by serial passages in vector-tick cell cultures. Ongoing studies have failed thus far.

In conclusion, this section has briefly described 12 years of research. It is intended to show that within this period the application of tick cell lines in arbovirus research has progressed from a very difficult beginning to the point where it can effectively be used as a valuable tool for the cultivation of certain arboviruses. Knowledge concerning arbovirus-vector-cell relationships could be considerably improved. According to results of studies concerned with the replication of viruses in tick cells, all enveloped tick-borne viruses grow in tick cells, but fail to do so in those of the mosquito *Aedes albopictus*. On the other hand, research in tick cell cultures and their use in arbovirus research has not attained advances achieved with other arthropod or vertebrate cell cultures. Before tick cell culture can be more successfully applied to studies in arbovirology, further improvements in tick cell culture methods and requirements must be achieved. This would stimulate additional investigations concerned with diagnostic problems of arbovirus infections, studies of pathogenesis, and studies of possible changes in virus characteristics by cultivation in tick cells.

II. EXPERIENCES WITH THE HANDLING OF TICK CELLS (LINE TTC-243) IN VITRO AND TESTS TO IMPROVE CELL ATTACHMENT AND GROWTH

The *Rhipicephalus appendiculatus* cell line TTC-243 was propagated and maintained by methods described by Varma et al.[3] and Pudney et al.[4] At first, cells were grown in Falcon® plastic flasks with a surface area of 25 cm² or in ordinary glass bottles; the cells were subcultured at a split ratio of 1:2 at 10-day intervals and incubated at 28 to 30°C. We now use a different plastic flask (Nunc) and notice better adherence and growth of seeded cells in these containers. However, alterations in morphology of tick cells cultured in different plastic or glass vessels were not seen. Usually, a monolayer of rounded cells mixed with a few spindle-like cells was formed within 7 to 10 days at 28 to 30°C. Higher temperatures were not tolerated by the cells. An increase of fetal bovine serum (FBS) from 10 to 20% resulted in an improved cell growth, as already observed by Kurtti and Munderloh.[2] Ultroser™ G is a fetal calf serum substitute suitable for in vitro cultures of numerous cell lines (Reactifs JBF, Soc. Chimique Pointed-Girard, France). The biological activity of 10 mℓ of a reconstituted solution is said to be equivalent to that of 50 mℓ of fetal calf serum. It would be of great value if Ultroser™ G could be used for tick cell media instead of fetal calf serum, as Ultroser™ G does not contain any antibody, is free of virus contamination, and is of constant quality.

Experiments to substitute FBS in L-15 medium by Ultroser™ G failed. Cells kept in a medium containing 4% Ultroser™ G (corresponding to 20% fetal calf serum) in L-15 medium grew significantly more slowly and monolayers were not formed in contrast to *Aedes albopictus, A. aegypti,* and *Toxorhynchites amboinensis* cells. We therefore discontinued use of Ultroser™ G for tick cell media.

It is very common to use "conditioned" (pretreated) tissue culture vessels for tick cell cultures in order to provide a better attachment of seeded cells on glass or plastic surfaces. Fibronectin (Fa. Seromed, Berlin) is derived from human plasma and, according to the information of the manufacturer, is able to increase the attachment of mammalian cells.

We investigated the influence of fibronectin on the adhesion of tick cells to plastic surfaces. Tested were 2 different solutions: (1) 0.15 mg fibronectin/mℓ medium; 2.5 mℓ of the mixture were added to each Nunc plastic flask which was kept for 30 min at room temperature; the solution was then washed off before seeding of the cell mixture; (2) 0.02 mg fibronectin were dissolved in 1 mℓ L-15 medium; 0.5 mℓ of the solution were added to each flask (25 cm^2) and distributed on the surface; the flasks were kept open at +37°C until the solution had dried off and then the cells were seeded. No significant improvement of the tick cell attachment was observed.

Very often we have noted that an extremely dense cell growth takes place at the rims and on the central areas of plastic flasks from different manufacturers. In such cases a uniform monolayer is not achieved. A useful manipulation may be "reseeding"[3] (although according to our experiences only at times); the cells were dispersed and allowed to reattach in the same flask. This can lead to a more even distribution and renewed vigor of the cells. We believe that differences in confluency between various tick cell lines may be treated in this way.

The role played by the inner surface of the culture vessels in influencing cell layer growth and density is very often so puzzling that we finally assumed electrostatic influences could be responsible — at least to some extent — for a nonhomogeneous cell growth. The incubators were therefore specially connected to the ground with copper wires, but without success. Apparently, minute chemical differences of various parts of the surface within distinct culture flasks cause such growth disturbances. We noticed that sometimes only single flasks of a batch are not suitable.

Attempts to revive tick cells from liquid N$_2$ storage were usually unsuccessful. We used DMSO (10%) and FBS (30%) in L-15 medium. The same technique is very effective when applied to mammalian cells.

III. PROPAGATION OF NAIROBI SHEEP DISEASE VIRUS IN THE TICK CELL LINE TTC-243

Details of NSDV (strain 045 Kenya) propagation in tick cells have been published.[13] Previously NSDV could be successfully passaged 45 times. From one culture flask, 3 to 4 virus-harvests could be made, after which the cells were "exhausted" and the amount of virus produced dropped by 2 to 3 logs$_{10}$. At this time the cells showed signs of degeneration. No further subcultures could be made. At no time could a typical and specific CPE be detected. The average virus amount/harvest was 7.5×10^6 LD$_{50}$. In contrast to expectations, no signs of attenuation of the virus strain used were observed. Virus of the 45th passage is as virulent for suckling mice as that of the first passages in tick cells. The incubation period for suckling mice inoculated intraperitoneally or intracerebrally remained at 5 to 6 days. No increase of titers in the mice was detected.

We conclude that tick cells (TTC-243) offer no striking advantages in comparison with mammalian cells (BHK, sheep kidney) for the multiplication of NSDV. Both systems provide about the same multiplication rate, but the cultivation and maintenance of tick cells is much more cumbersome, time consuming, and expensive. In addition, more experience is necessary than for the handling of mammalian cells.

Infected tick cells do not show CPE. Virus multiplication must therefore be determined in other sensitive host systems, preferably suckling mice. This considerably hampers the use of tick cells in the detection of NSDV in field samples. Therefore tick cells are presently useful only for special purposes. It has not yet been possible to show whether NSDV strain 045 adapted to tick cells (45 passages) is of reduced virulence to European sheep in comparison to the original field isolates.

IV. DETECTION OF NAIROBI SHEEP DISEASE VIRUS ANTIGEN BY IMMUNOPEROXIDASE IN THE TICK CELL LINE TTC-243

NSDV propagation in tissue culture is not without its problems. Sheep kidneys which are used in the preparation of cell cultures are not readily available everywhere and, as with tick cells, some BHK-cell lines do not indicate virus multiplication by CPE. Detecting virus multiplication in infected tissue culture systems in the absence of CPE is possible by two practical means, the demonstration of virus antigen by immunofluorescence (IF) and by immunoenzymatic methods.

Recently we reported successful detection of NSDV antigen by direct and/or indirect immunoperoxidase (IP) techniques in infected sheep kidney and BHK-13 cells and in the TTC-243 cell line, respectively.[14] In the indirect technique the medium was removed carefully and the monolayers were washed twice with cool phosphate-buffered saline (PBS) and fixed in −20°C cold acetone (80%) for 15 min at room temperature. No negative influence on the surfaces of plastic materials was observed and the cell sheets were well stabilized for subsequent treatment. After removing the acetone and drying the cells in an air stream, we added 15 mℓ immune serum. The flasks were incubated at 37°C for 1 hr and then washed 3 times with PBS. Five mℓ of conjugate, rabbit-antisheep immunoglobulin G (H + L) conjugated with peroxidase (Inst. Pasteur Prod., Paris, France) or a rabbit-antisheep immunoglobulin G (F[ab′]2), fragment-specific conjugated with peroxidase (Cappel Lab., Malvern, Penn.) diluted in PBS were added, followed by further incubation at 37°C. The reaction was microscopically observed and halted with tap water as soon as a brown staining of the cytoplasm of cells could be observed.

By indirect IP tests, two colored cell-types could be differentiated: one was spherical or droplet-like and completely of a uniform brown color and the other was spindle-shaped with many dark brown granules in the cytoplasm, but without a colored nucleus (Figure 1). Best results were obtained between 3 and 6 days after inoculation and not before, but the test seemed to be less sensitive in tick cells than in BHK or sheep kidney cells.

Improved IP techniques are an important aid in the detection of virus antigens and their antibodies and in the study of virus replication at both optical and electron microscopic levels in infected cells. These techniques are similar in sensitivity to IF tests and also offer operational advantages.

IP techniques have also been used in arbovirus research where it has been possible to demonstrate the replication, processing, and localization of several species of Togaviridae in arthropods and invertebrate in vitro systems. Different mechanisms in the replication between alpha- and flaviviruses were demonstrated.[15]

We adapted known principles of IP techniques to demonstrate NSDV antigen qualitatively in *Rhipicephalus appendiculatus* cells (TTC-243) by indirect IP tests. The specificity of the reaction described is supported by the fact that the sera used produced positive results in ELISA (enzyme-linked immunosorbent assay) and in indirect IF and hemagglutination tests as well.[16]

Optimal reactions largely depend on the quality of the plastic culture vessels used and on the density of the infected cell sheet. Premature detachment of cells is likely if too many cells are seeded, whereas an insufficient growth of cells occurs if too few cells are used.

Nonspecific intensive coloring occurs in dead, uninfected cells after mechanical or

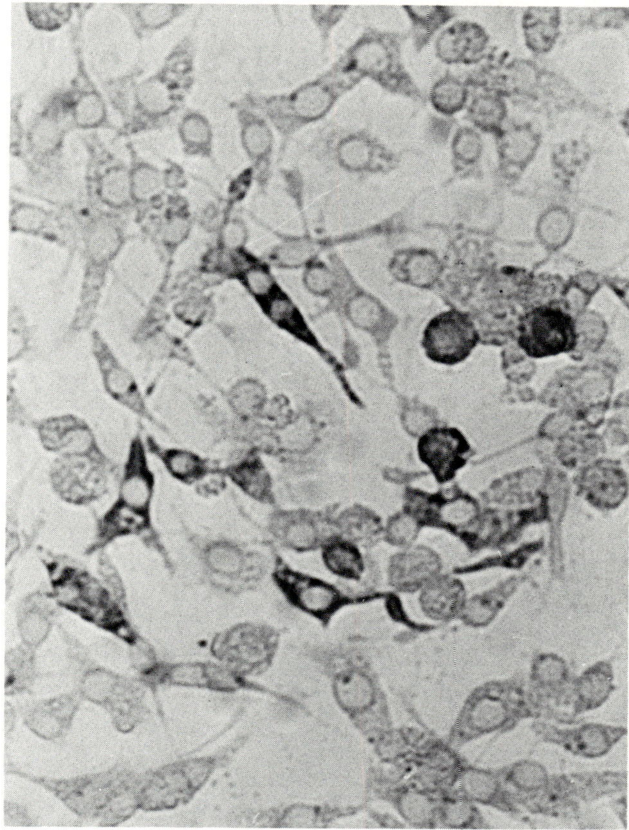

FIGURE 1.   Positive immunoperoxidase-reaction in cultivated and NSDV-infected tick cells (line TTC-243).

chemical damage. This is much more frequently observed with tick cells than with BHK or sheep kidney cells. Tick cells need a considerably longer time to form monolayers. During this time many cells degenerate and die, but remain attached to the monolayer. Such "false positive" staining may result in a misinterpretation. Another similar difficulty sometimes arises with a very dense monolayer where cells tend to have a very small amount of cytoplasm that may become brownish. In such cases it is difficult to differentiate these possibly uninfected cells from specifically staining cells. This emphasizes the importance of using monolayers of optimal confluency.

The application of the IP technique for localization of NSDV antigen in the cytoplasm of infected cells at light microscopic levels does not reveal inclusion-like bodies which are demonstrable in mammalian cells stained with hematoxylineosin.[17] A positive reaction in tick cells is not strictly located in distinct areas of the cytoplasm. This apparently depends on the morphology of the cultured tick cells. A positive reaction in BHK or sheep kidney cells is indicated by a distinct granular staining of the cytoplasm usually in the form of a ring or semicircle around the unstained nucleus, with many parts of the cytoplasm being spared. This picture corresponds more to that of inclusion bodies as seen after staining with hematoxylineosin.

In the preceding section we have discussed the advantages and disadvantages of the indirect IP technique in detecting NSDV antigen in tick cell cultures (TTC-243). Positive IP reactions were observed in the absence of a virus-induced CPE. The indirect IP technique is suitable for detection of NSDV infection in tick cells and therefore may adequately replace mouse-inoculation tests. However, critical evaluation leads at present to the conclusion that tick cells are less suitable for such purposes than susceptible mammalian cells.

V. DETECTION OF NAIROBI SHEEP DISEASE VIRUS AND CONTAMINATION WITH *REOVIRUS*-LIKE PARTICLES IN TICK CELLS BY TRANSMISSION ELECTRON MICROSCOPY

Detailed ultrahistological studies on tick tissues, primary cultures of tick cells, or tick cell lines are very scarce and only a very few investigations dealing with arbovirus-infected ticks[18] of tick cells cultivated in vitro have been published. This is in contrast to a fairly great number of publications concerned with research on virus-infected mosquitoes or cultivated mosquito cells.

Reháček[19] investigated by electron microscopy primary cell cultures prepared from developing adults of *Rhipicephalus sanguineus* ticks. He found that in 1-week old cultures such cells do not possess special arrangements of organelles. In general, no differences in ultrastructure from in vitro cultures of other arthropod cells were detectable. Inclusions, viral particles, or other pathogens could not be demonstrated in any of the cells examined (unfortunately photographs were not included).

Recently we reported electron microscopic studies of the morphology of NSDV.[20] Extracellular and negatively stained NSDV particles have a spherical shape which does not seem to be very stable. Their average diameter was determined to be 110 nm, but variations from 90 to 120 nm were seen. The surface of NSDV is covered with projections of 10 nm in length. Below this sheet of projections we sometimes were able to observe a membrane of lesser electron density of approximately 5 nm thickness. The inner part of the particles is electron-dense to various degrees and therefore appears brighter or darker. The structure of this "core" appears to be homogeneous or more or less granular.

Infected sheep kidney cells show many intracytoplasmic vacuoles. In such vacuoles we found spherical particles either individually or in groups of up to 20. Their morphology was very similar to those of other bunyaviruses or nairoviruses, respectively.[21] All were round or oval, limited by a 9.5 nm thick layer which showed a radial structure, and followed by a membrane-like layer of 4 nm thickness. The inner part of the particles appeared to be granular and of different electron density without regular arrangement. The average diameter of the particles was approximately 80 nm.

We have now extended these studies by ultrahistological investigations of noninfected tick cells (line TTC-243) and cells of this line infected with NSDV.

Monolayers of noninfected tick cells were fixed *in situ* on the surface of the plastic culture vessels with 2% glutaraldehyde in a buffer at pH 7.2. After washing of the cell sheet with buffer, a second fixation was carried out with 2% osmium tetroxide and further washings followed. The cell sheet was then scraped off and centrifuged. Dehydration was achieved by ethanol treatment. Polymerization was carried out in Durcopan ACM (Fluca). We used an Ultratome (LKB) to prepare 40 nm thick sections which were stained with uranylacetate (in 70% ethanol) and Reynolds lead citrate. We worked with a Zeiss EM C/R transmission electron microscope and formvar or pioloform-coated grids at 60,000 V. NSDV-infected tick cells harvested at 24, 30, 48, and 52 hr postinoculation were treated identically.

Striking differences were found in the morphology between tissue culture cells of vertebrates and cultured tick cells. The latter presented a very heterogeneous cytoplasm characterized by many unequally distributed, rounded, and electron-dense lipid structures measuring about 50 to 300 nm by amorphous and electron-dense material con-

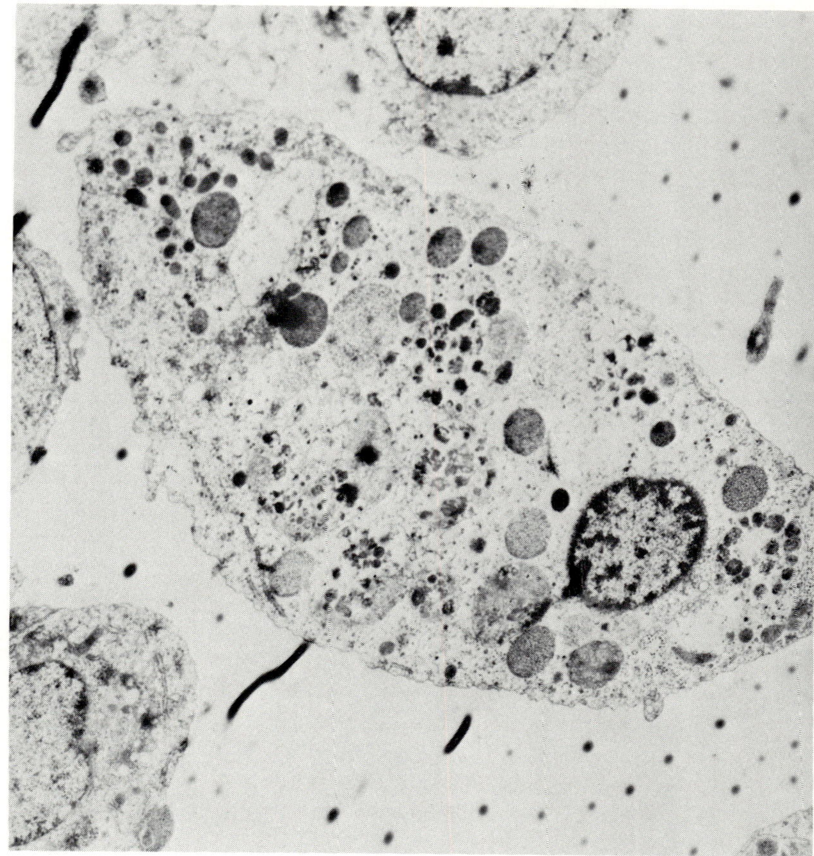

FIGURE 2

FIGURES 2 AND 3.   In vitro cultivated tick cells (line TTC-243) (magnification × 10,000).

centrated in small centers and by many membrane-like structures. Intracytoplasmic vacuoles were frequently found. They were partially filled with electron-dense material. Mitochondria were common, but endoplasmatic reticulum was not often seen. No differences were found in the morphological appearance of the nuclei of vertebrate and tick cells (Figures 2 and 3). The numerous electron-dense rounded or multiform structures, often in the size range of viruses, have rendered the search for NSDV very difficult.

A. Demonstration of *Reovirus*-Like Particles

Reovirus-like particles were found in noninfected and NSDV-infected tick cells in small areas of the cytoplasm near the nucleus. Surprisingly, these particles were seen more often and more numerously in infected than in noninfected cells, where they were found only occasionally, as though an infection of the tick cells with NSDV would trigger an increased formation of the particles. The *Reovirus*-like particles measured 50 to 55 nm in diameter, were round or polygonal, and had an electron-dense core surrounded by a ring-like structure which could be identical with a capsid. The majority of the particles were embedded in granulated zones of the cytoplasm or in areas with granulated thread-like material (Figures 4 to 6). No specific or nonspecific alterations of tick cells containing *Reovirus*-like particles were observed.

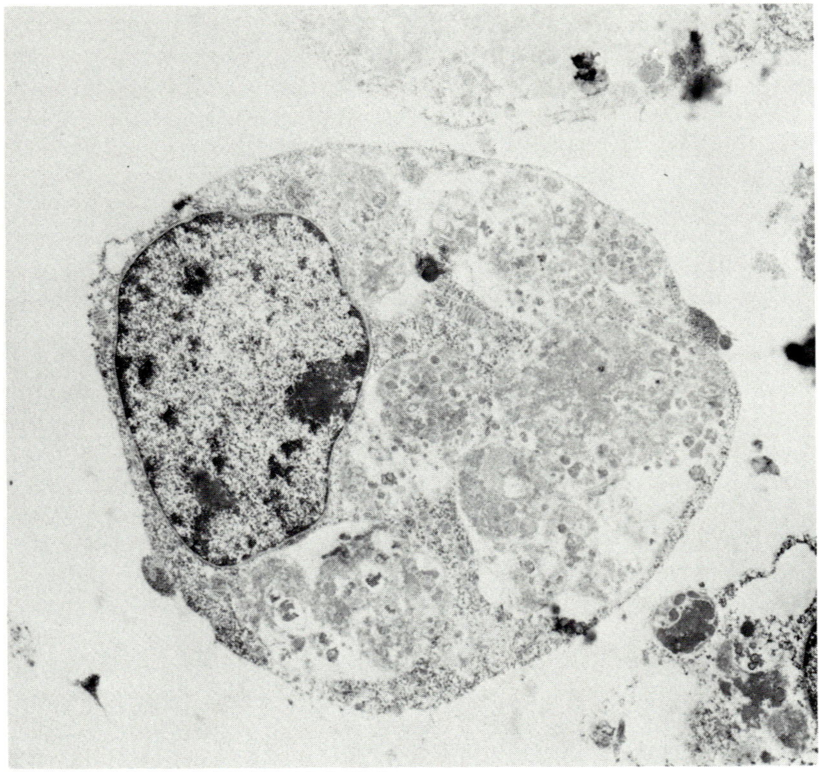

FIGURE 3

B. Demonstration of NSDV in Tick Cells

In NSDV-infected tick cells we found virions only infrequently, mostly in small groups within homogeneous, apparently unaltered cytoplasm (Figure 7). They were round or polygonal and their average diameter was 90 to 100 nm. The core of the particles was partially electron-dense. Spike-like structures on their surface, as seen in infected sheep kidney cells, could not be identified.

On the whole the morphology of the particles strongly resembled that found in NSDV-infected sheep kidney cells[24] and those virions which are regarded as bunyaviruses or nairoviruses, respectively.[21,22] In noninfected tick cells such virions could not be detected.

Virus contamination of arthropod cell lines mostly of *Aedes* and *Drosophila* origin has been reported since 1974 rather frequently.[5,23,24] As far as we know viral contamination of a tick cell line has not been reported yet, except for a remark by Pudney et al.[12] that virus-like particles had been observed in a *Rhipicephalus appendiculatus* cell line. Here we present the first evidence of a virus contamination of tick cells cultivated in vitro. The ultrahistological examination of normal, noninfected tick cells revealed *Reovirus*-like particles measuring 50 to 55 nm. At times these showed a regular arrangement. The virus contamination did not cause CPE or any other cell alterations. Experiments to isolate these *Reovirus*-like particles for further identification is planned.

Reoviruses have been found in *Drosophila* cell lines[23,24] and it has been suggested that these viruses are inherited. On the other hand, viral contamination of *Drosophila* cells has also been suggested to come from FBS used in the medium.[23]

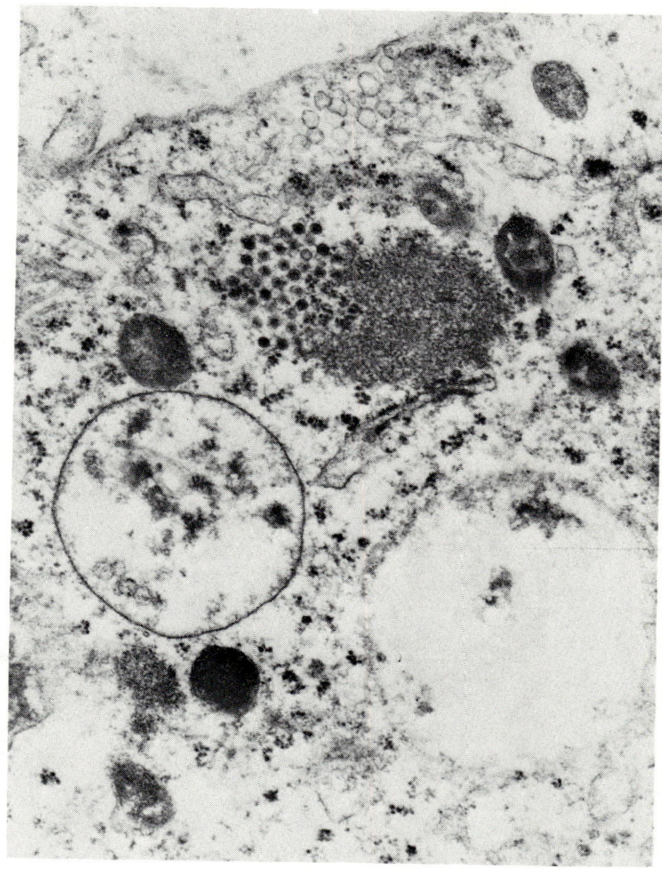

FIGURE 4

FIGURES 4 THROUGH 6. Reovirus-like particles in cultivated tick
cells (line TTC-243) (magnification × 30,000).

Since the origin of the *Reovirus*-like particles in our tick cells is unknown, it should
be noted that the batches of FBS used for tick cell culture medium are the same as for
all other mammalian cell cultures kept in our laboratory and no signs of viral contamination in these cells have been found. It is therefore assumed that the particles were
nonvirulent and inherent in viruses of the tick cell line TTC-243 and not of extraneous
origin.

At present there is an increasing number of reports dealing with virus-infected ixodid
and argasid ticks. Different arboviruses could be detected, mostly species of the family
Bunyaviridae, Togaviridae, and Reoviridae (genus *Orbivirus*), and also probably *Or-
thomyxovirus* (see Arthropod-borne Virus Information Exchange, 1983 to 1984).

The observation of a viral contamination underlines the importance of carefully
monitoring tick cell cultures for unexpected viral agents by electron microscopic examination, as obvious visible effects on the contaminated cells may not be apparent.
The possibility of viral contamination of tick cells must be kept in mind at all times,
particularly when tick cells are used for virological studies.

Only a few particles of NSDV could be detected in TTC-243 cells. These were mostly
located within the cytoplasm, but some were found between cells in the intercellular

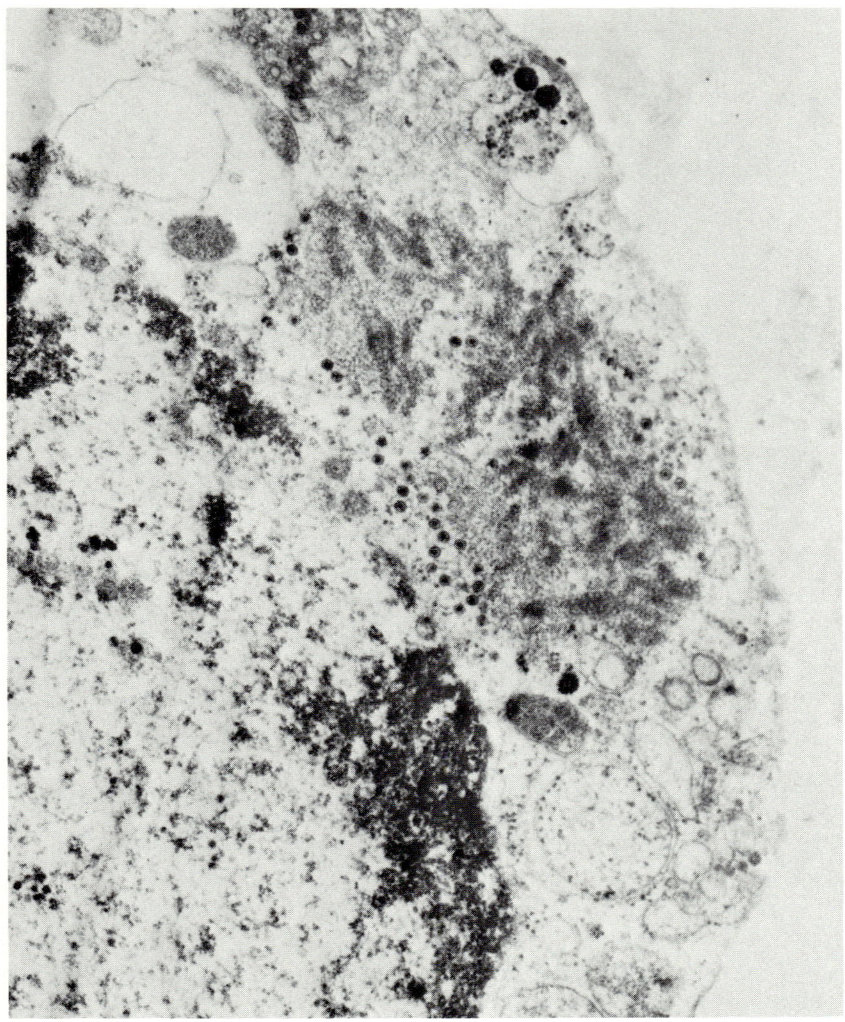

FIGURE 5

spaces. The size of the particles was approximately 90 to 100 nm mean diameter and showed a double membrane. The morphology of these particles corresponds very well with those which we found in the brains of intracerebrally infected suckling mice (unpublished data) and sheep kidney cells,[21] and shows a close similarity with that of other bunyaviruses. As the titers of infected tick cells are usually not higher than $10^{5.5}$ LD$_{50}$/ mℓ, electron microscopic examination of inoculated tick cell cultures is not a sensitive method of detecting NSDV because this method requires titers of at least 5.0 log 10 LD$_{50}$ or ID$_{50}$/mℓ to insure virus detection.

One electron microscopic report concerned with the demonstration of Soldado virus (genus *Nairovirus*) in the tissue of an *Ornithodorus (Alectorobius) maritimus* tick has recently been published by Chastel et al.[25] Clusters of the virus were infrequent and found in pseudovacuoles located in an amorphous material bordering the plasma membrane of unidentified cells belonging possibly to a salivary gland acinus. Virus particles were round and pleomorphic with a mean diameter of 102 nm (88 to 120 nm) and limited by a trilamellar envelope without obvious "spikes". The internal density of

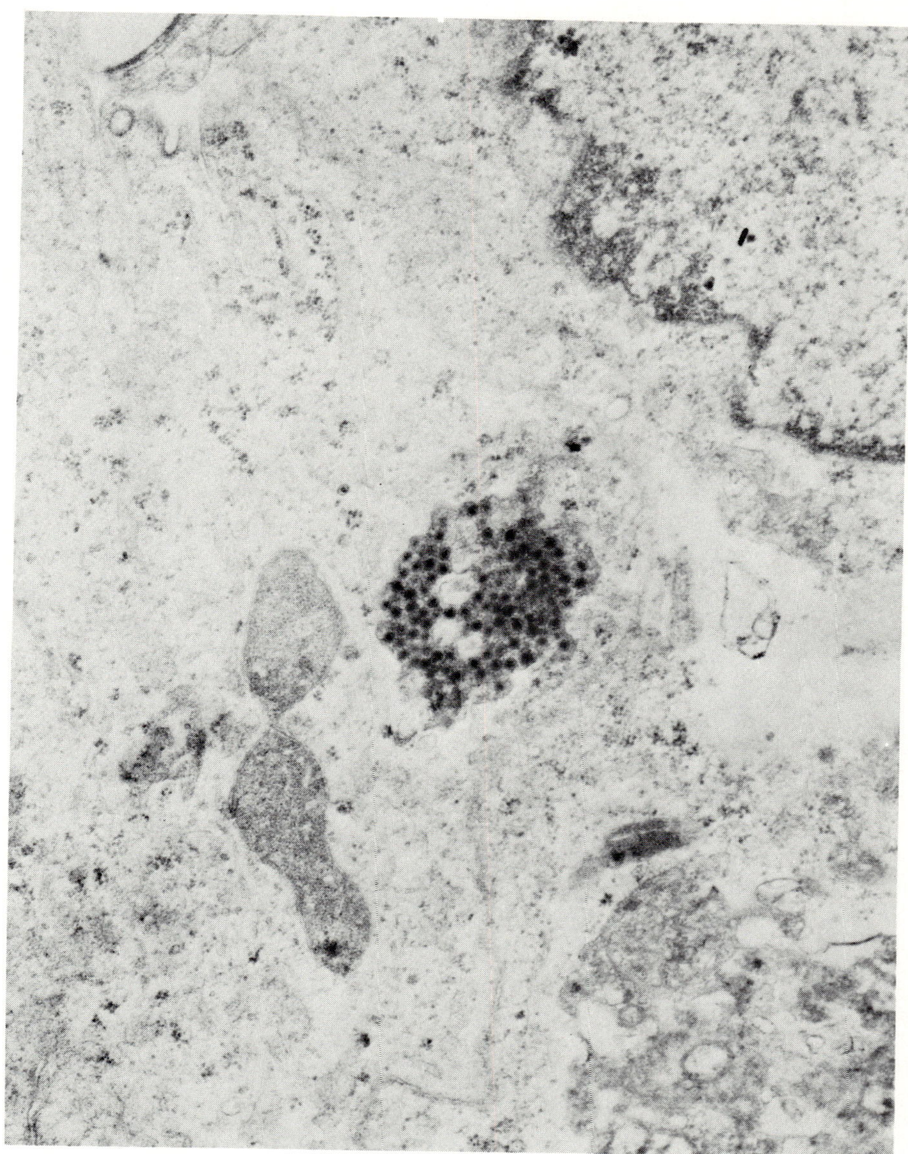

FIGURE 6

particles was reinforced beneath the envelope. The authors emphasize that the actual classification of certain particles accidentally found by transmission electron microscopy in naturally infected ticks is frequently difficult to establish. Another *Nairovirus* (Uukuniemi subgroup) also showed a typical *Bunyavirus* morphology in sectioned mouse brains infected with material of *Ixodes uriae* ticks.[26] Our results are in close agreement with these observations. We therefore conclude that NSDV has been visualized for the first time in tick cells cultured in vitro.

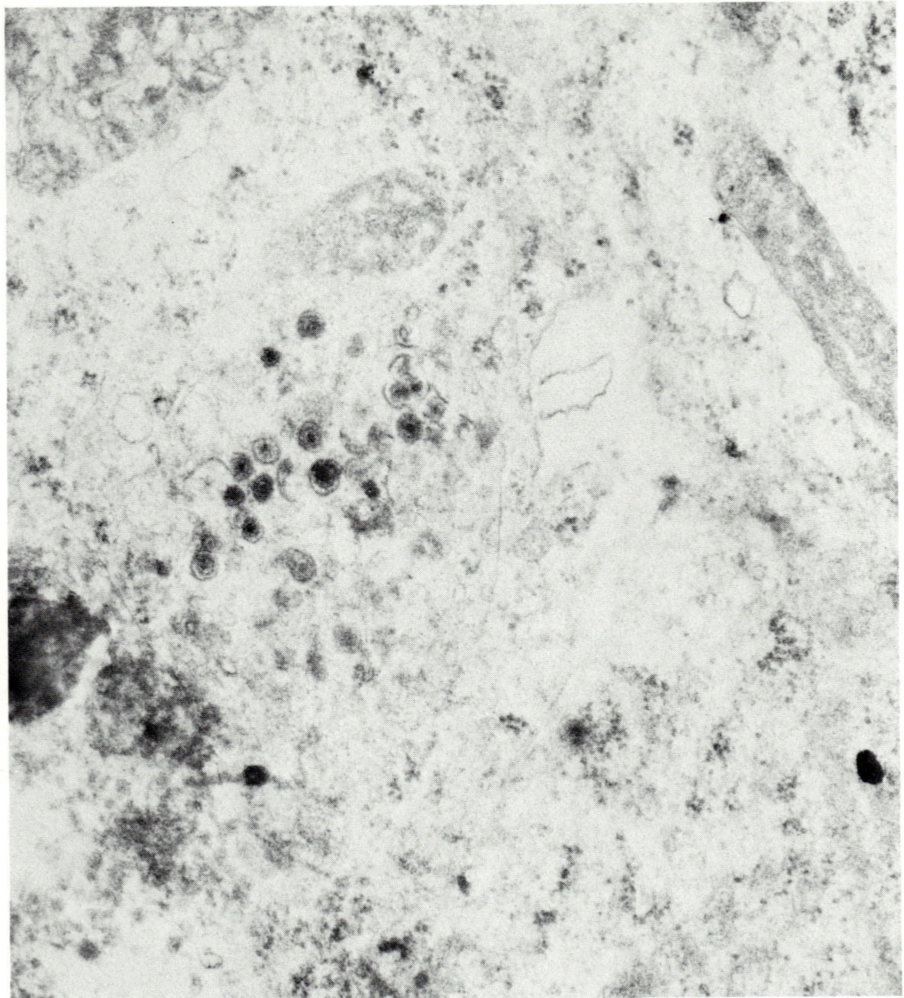

FIGURE 7.   Nairobi Sheep Disease Virus particles in cultivated tick cells (line TTC-243) (magnification × 48,000).

REFERENCES

1. Kurtti, T. J. and Büscher, G., Trends in tick cell culture, in *Practical Tissue Culture Applications,* Maramorosch, K. and Hirumi, H., Eds., Academic Press, New York, 1979, 351.
2. Kurtti, T. J. and Munderloh, U. G., Tick cell culture: characteristics, growth requirements, and applications to parasitology, in *Invertebrate Cell Culture Applications,* Maramorosch, K. and Mitsuhashi, J., Eds., Academic Press, New York, 1982, 195.
3. Varma, M. G. R., Pudney, M., and Leake, C. J., The establishment of three cell lines from the tick *Rhipicephalus appendiculatus* (Acari: Ixodidae) and their infection with some arboviruses, *J. Med. Entomol.,* 11, 698, 1975.
4. Pudney, M., Varma, M. G. R., and Leake, C. J., The growth of some arboviruses in tick cell lines, in *Tick-borne Diseases and Their Vectors,* Wilde, J. K. H., Ed., University of Edinburgh, 1978, 490.
5. Pudney, M., Leake, C. J., and Varma, M. G. R., Replication of arboviruses in arthropod in vitro systems, in *Arctic and Tropical Arboviruses,* Kurstak, E., Ed., Academic Press, New York, 1979, 245.

6. David-West, T. S., Propagation and plaquing of Dugbe virus (an ungrouped Nigerian arbovirus) in various mammalian and arthropod cell lines, *Arch. Gesamte Virusforsch.*, 44, 330, 1974.

7. Pudney, M., Varma, M. G. R., and Leake, C. J., Culture of embryonic cells from the tick *Boophilus microplus* (Ixodidae), *J. Med. Entomol.*, 10, 493, 1973.

8. Leake, C. J., Pudney, M., and Varma, M. G. R., Studies on arboviruses in established tick cell lines, in *Invertebrate Systems In Vitro*, Kurstak, E., Maramorosch, K., and Dübendorfer, A., Eds., Elsevier/North-Holland, Amsterdam, 1980, 327.

9. Bhat, U. K. M. and Yunker, C. E., Susceptibility of a tick cell line (*Dermacentor parumapertus* Neumann) to infection with arboviruses, in *Arctic and Tropical Arboviruses*, Kurstak, E., Ed., Academic Press, New York, 1979, 263.

10. Buckley, S. M., Hayes, C. G., Maloney, J. M., Lipman, M., Aitken, T. H. G., and Casals, J., Arbovirus studies in invertebrate cell lines, in *Invertebrate Tissue Culture*, Kurstak, E. and Maramorosch, K., Eds., Academic Press, New York, 1976, 3.

11. Hink, W. F., The 1979 compilation of invertebrate cell lines and culture media, in *Invertebrate Systems In Vitro*, Kurstak, E., Maramorosch, K., and Dübendorfer, A., Eds., Elsevier/North-Holland, Amsterdam, 1980, 553.

12. Pudney, M., Leake, C. J., and Buckley, S. M., Replication of arboviruses in arthropod in vitro systems, in *Invertebrate Cell Culture Applications*, Maramorosch, K. and Mitsuhashi, J., Eds., Academic Press, New York, 1982, 159.

13. Munz, E., Reimann, M., Munderloh, U., and Settele, U., The susceptibility of the tick cell line TTC-243 for Nairobi Sheep Disease virus and some other important species of mammalian RNA and DNA viruses, in *Invertebrate Systems In Vitro*, Kurstak, E., Maramorosch, K., and Dübendorfer, A., Eds., Elsevier/North-Holland, Amsterdam, 1980, 337.

14. Munz, E., Reimann, M., and Jäger, H., Qualitative and quantitative detection of Nairobi sheep disease virus antigen by immunoperoxidase in BHK-, sheep kidney- and tick cell cultures, *Zentralbl. Veterinaermed. Beih.*, 31, 231, 1984.

15. Kurstak, E., Tijssen, P., van den Hurk, J., Kurstak, C., and Morisset, R., Detection by immunoperoxidase and ELISA of arbovirus antigens and antibodies in in vivo and in vitro systems, in *Invertebrate Systems In Vitro*, Kurstak, E., Maramorosch, K., and Dübendorfer, A., Eds., Elsevier/North-Holland, Amsterdam, 1980, 365.

16. Munz, E., Reimann, M., Fritz, T., and Meier, K., An enzyme-linked immunosorbent assay (ELISA) for the detection of antibodies to Nairobi sheep disease virus in comparison with an indirect immunofluorescent — and haemagglutination test. II. Results observed with sera of experimentally infected rabbits and sheep and with African sheep sera, *Zentralbl. Veterinaermed. Beih.*, 31, 537, 1984.

17. Coackley, W. and Pini, A., The effect of Nairobi sheep disease virus on tissue culture systems, *J. Pathol. Bacteriol.*, 90, 672, 1965.

18. Nosek, J., Korolev, M. B., Chunikhin, S. P., Kožuch, O., and Čiampor, F., The replication and eclipse-phase of the tick-borne encephalitis virus in Dermacentor reticulatus, *Folia Parasitol. (Prague)*, 31, 187, 1984.

19. Reháček, J., Tick tissue culture and arboviruses, in *Invertebrate Tissue Culture Applications in Medicine, Biology and Agriculture*, Kurstak, E. and Maramorosch, K., Eds., Academic Press, New York, 1976, 21.

20. Munz, E., Göbel, E., Krolopp, C., Reimann, M., and Davies, F. G., Electron microscopic studies on the morphology of Nairobi sheep disease virus, *Zentralbl. Veterinaermed. Beih.*, 28, 553, 1981.

21. Clerx, J. P. M., Casals, J., and Bishop, D. H. L., Structural characteristics of Nairoviruses (genus *Nairovirus*, Bunyaviridae), *J. Gen. Virol.*, 55, 165, 1981.

22. Murphy, F. A., Harrison, A. K., and Whitfield, S. G., Bunyaviridae: morphologic and morphogenetic similarities of Bunyamwera serologic supergroup viruses and several other arthropod-borne viruses, *Intervirology*, 1, 297, 1973.

23. Plus, N., Further studies on the origin of the endogenous viruses of Drosophila melanogaster cell lines, in *Invertebrate Systems In Vitro*, Kurstak, E., Maramorosch, K., and Dübendorfer, A., Eds., Elsevier/North-Holland, Amsterdam, 1980, 327.

24. Gateff, E., Gissmann, L., Shrestha, R., Plus, N., Pfister, H., Schröder, J., and zur Hausen, H., Characterization of two tumorous blood cell lines of Drosophila melanogaster and the viruses they contain, in *Invertebrate Systems In Vitro*, Kurstak, E., Maramorosch, K., and Dübendorfer, A., Eds., Elsevier/North-Holland, Amsterdam, 1980, 517.

25. Chastel, C., Guiguen, C., Quillien, M. C., Le Lay-Roguès, G., and Beaucornu, J. C., Visualization of Soldado virus by electron microscopy in the tissue of the vector tick Ornithodorus (Alectorobius) maritimus Vermeil and Marguet, 1967, *Ann. Parasitol.*, 59, 1, 1984.

26. Eley, S. M. and Nuttall, P. A., Isolation of an english Uukuvirus (family Bunyaviridae), *J. Hyg. Camb.*, 93, 313, 1984.

Index

INDEX

A

D

E

F

G

Trypsin, 26, 32, 47
Trypsinization, 41
Typhus rickettsie, 36
Tyrosinase, 27

U

UUK, see Uukuniemi virus
Uukuniemi (UUK) virus, 97, 145

V

Vaccine and tick cells, 96
Vaccinia virus, 94
Venezuelan equine encephalitis (VEE), 12, 68, 69
Versene, 47
Vertebrate blood vs. insect hemolymph, 6
Vertebrate cells
  arthropod cells vs., 8
  tick cells vs., 140—141
Vitamin B, 28

W

Washing, 72
WEE, see Western equine encephalitis

Western equine encephalitis (WEE), 4, 5, 7, 68, 69
  interference with, 18
West Nile (WN) virus, 4—5, 7, 68, 69, 73, 90
  in tick cells, 13
WHA, see Whataroa virus
Whataroa (WHA) virus, 89
WN, see West Nile virus

X

Xanthine dehydrogenase (XDH), 59, 60
XDH, see Xanthine dehydrogenase
Xenopus laevis, 89, 97

Y

Yellow fever (YF), 80, 84—85, 90
  Asibi strain of, 5
  monoclonal antibodies in, 98
Yellow fever (YF)-IgM immunocomplexes, 85
YF, see Yellow fever

Z

ZIR, see Zirqa virus
Zirqa (ZIR) virus, 88, 93—95, 134